Terrain and Tactics

Recent Titles in
Contributions in Military Studies

The Army's Nuclear Power Program: The Evolution of the Support Agency
Lawrence H. Suid

Russian Imperial Military Doctrine and Education, 1832–1914
Carl Van Dyke

The Eagle and the Dragon: The United States Military in China, 1901–1937
Dennis L. Noble

An Eyewitness Account of the American Revolution and New England Life: The Journal of J. F. Wasmus, German Company Surgeon, 1776–1783
Helga Doblin, translator, and Mary C. Lynn, editor

The Declining Hegemon: The United States and European Defense, 1960–1990
Joseph Lepgold

Diplomatic Ramifications of Unrestricted Submarine Warfare, 1939–1941
Janet M. Manson

From Sumer to Rome: The Military Capabilities of Ancient Armies
Richard A. Gabriel and Karen S. Metz

The Political Economy of Defense: Issues and Perspectives
Andrew L. Ross, editor

Left Face: Soldier Unions and Resistance Movements in Modern Armies
David Cortright and Max Watts

Mechanized Juggernaut of Military Anachronism?: Horses and the German Army of World War II
R. L. DiNardo

Militarism and Politics in Latin America: Peru from Sánchez Cerro to *Sendero Luminoso*
Daniel M. Masterson

Salzburg Under Siege: U.S. Occupation, 1945–1955
Donald R. Whitnah and Florentine E. Whitnah

TERRAIN A·N·D TACTICS

PATRICK O'SULLIVAN

Contributions in Military Studies, Number 115

GREENWOOD PRESS

New York • Westport, Connecticut • London

Library of Congress Cataloging-in-Publication Data

O'Sullivan, Patrick.
Terrain and tactics / Patrick O'Sullivan.
p. cm.—(Contributions in military studies, ISSN 0883–6884 ; no. 115)
Includes bibliographical references and index.
ISBN 0–313–27923–3 (alk. paper)
1. Military geography. 2. Tactics. I. Title. II. Series.
UA990.079 1991
355.4'7—dc20 91–12223

British Library Cataloguing in Publication Data is available.

Library of Congress Catalog Card Number: 91–12223
ISBN: 0–313–27923–3
ISSN: 0883–6884

First published in 1991

Greenwood Press, 88 Post Road West, Westport, CT 06881
An imprint of Greenwood Publishing Group, Inc.

Printed in the United States of America

The paper used in this book complies with the Permanent Paper Standard issued by the National Information Standards Organization (Z39.48–1984).

10 9 8 7 6 5 4 3 2 1

CONTENTS

	Figures and Tables	vii
1.	Military Geography	1
2.	Some Third World Wars	5
3.	The Lie of the Land	31
4.	The World at War	41
5.	The Geography of Revolution	71
6.	The Geography of Battles	87
7.	Classic Spatial Ploys	105
8.	Terrain and Tactics	113
9.	Guerrillas and Counterinsurgency	125
10.	War in Cities	137
11.	Northern Ireland	149
12.	Fighting in the Landscape and Fighting for a Place	163
	Bibliography	169
	Index	175

FIGURES AND TABLES

FIGURES

2.1 The Western Sahara 10
2.2 Iran-Iraq 12
2.3 Ogaden 14
2.4 Lebanon 16
2.5 Afghanistan 17
2.6 The Falklands 19
2.7 North Vietnam 20
2.8 Central America 22
4.1 War in the Late 1940s 42
4.2 War in the 1950s 43
4.3 War in the 1960s 45
4.4 War in the 1970s 46
4.5 War in the 1980s 47
4.6 Power Gradients 50
5.1 The Power Pyramid 81
5.2 Peripheral and Internal Rebel Positions 82
5.3 Rebels with Steeper Power Gradients 83
5.4 Rebels Take the Center 84

9.1 Guerrilla Success, Rain and Slope 132
10.1 Tactical Widths and Line-of-Sight Distances 144
11.1 Incidents Involving the British Army in 1974 154

TABLES

9.1 Population Densities and Regular-to-Guerrilla Ratios 128
9.2 Guerrilla Campaigns Since 1945 129
10.1 Fighting Sector Widths 142
10.2 Line-of-Sight Distances 143
10.3 Line-of-Sight Distributions for Dispersed Residential Areas 145

1

MILITARY GEOGRAPHY

War is the usual and all too frequent means of resolving political conflicts. Wars are almost invariably about who controls what territory. Whether the fight be civil or between nations, it usually involves turf and, thus, geography. While political geography addresses the causes of these conflicts, military geography consists of the use of geographical knowledge in describing and analyzing the employment of armed force. Although geographers have provided advice and expertise for war, the object of this book is not to tell soldiers how to win. The attitude adopted is analytic and academic rather than prescriptive and partisan. The objective is to weigh the balance of advantage for combatants in different geographic settings, taking as material and impersonal a viewpoint as possible. The moral hope underlying this endeavor is that war will prove futile, that the benefit of the outcome will prove to be swamped by the enormous cost in lives and resources inevitably involved. It is toward those measurements and judgments that what is offered here contributes.

The geographer's viewpoint concentrates attention on the significance of location, distance, landscape and climate in determining the outcome of war. There is, of course, a long history of the application of geographical skills to military purposes. Much geographical exploration was done for this purpose. Topographic mapmaking emerged to serve artillery needs. The foundation of descriptive, regional geography was laid by naval intelligence services. The geographer's chief tool, the map, remains the premier instrument of military intelligence, decision and command. Intelligence on geographical conditions is fundamental to tactical or strategic success. Field Marshal Montgomery credited victory in battle to "transportation, administration and geography."

Besides this service of military needs, geographers have described in a scholarly vein the foundations and disposition of military might and its impact on society and the landscape. The attention of this book is a little more immediate, being concerned with the actual fighting. Here we will explore the effect of geographical circumstances on the outcome of violent conflict. This could obviously be done in a particular, historical fashion, looking at some specific campaigns as case studies. There is a stage beyond, using the stuff of history as raw material for the construction of some general principles. These then are first tottering steps in that direction.

To take off and get into the thick of it as fast as possible, we will seek what lessons can be learned from a selection of recent wars about the effects of global position, environmental conditions and geographical detail on the interplay of geostrategy, tactical decisions and results. Having used these to generate a motif, we next build the setting for a discussion of the central theme of the book, the relationship between tactics and terrain. First, the earth's variety of physical habitats and the pattern of our occupation of them are surveyed from a military viewpoint.

To this is added a geopolitical veneer, consisting of a geography of war since 1945 and a survey of the present sources of conflict, actual and potential. Since so much strife is contained within states, involving efforts to overthrow the existing political structures, it is necessary to consider the geography of revolution, as well as the geography of international territorial competition.

The raw stuff to fuel a discourse on tactics is provided by a geographical analysis of a selection of battles, ancient and modern. Out of this comes a characterization of classical tactical ploys. This provides the other component for the calculation of the tactical balance in the diversity of settings where war takes place. Following a broad examination of modern weapons, tactics and the required appreciation of the battlefield, one particular form of warfare and one particular environment are deemed worthy of more thorough treatment. What is euphemistically referred to as "low-intensity warfare" is considered in an analysis of guerrillas and counterinsurgency operations. Then, since the streets and buildings of towns are increasingly the venue for fighting, urban terrain and its salient features are examined from a tactical viewpoint. To draw the strands of political geography, strategy and tactics together with a sustained example, this is followed by a description of the twenty-year-old contest between the British Army and the Irish Republican Army (IRA), which includes both rural and urban settings.

Having explored geography from the perspectives of politicians and commanders, those who cause and command wars, it seems appropriate to finish by looking at the perceptions of those who do the actual killing and dying. We end with the geographical aptitudes and attitudes of soldiers, the fundamental particles of tactics and, thus, of strategy and geopolitics.

REFERENCES

O'Sullivan, P., and J. Miller. 1983. *The Geography of Warfare*. London: Croom Helm.

Peltier, L. C., and G. E. Pearcy. 1966. *Military Geography*. Princeton, N.J.: Van Nostrand.

2

SOME THIRD WORLD WARS

In 1985 I was invited to write an essay on the geography of a selection of Third World wars for a book entitled *The Lessons of Recent Wars in the Third World* (Neuman and Harkavy 1987). The editors had collected analyses of the conflicts in Afghanistan (1979–89), Central America (1980s), the Falklands (1982), Iran and Iraq (1980–88), Lebanon (1982), North Vietnam (1979), Ogaden (1977–78) and the Western Sahara (1975–87). To these they were adding general commentaries on the cultural, sociological, economic and political aspects of these wars, and they felt the need to treat their geographical circumstances. This proved to be an interesting exercise because the wars in question were fought over a range of very different physical and human landscapes, and they stemmed from diverse geopolitical conflicts. The array of global, political and environmental settings involved had generated a variety of strategic, operational and tactical responses; it included everything from guerrillas in the rain forest of Central America to a national confrontation on the dry plains of Mesopotamia. In weaponry it ran the gamut from electronic air and naval warfare in the South Atlantic to Pathan ambushes with handmade copies of Lee Enfield 303s or stolen Kalashnikovs in the Hindu Kush.

Afghanistan was invaded by the Soviet Army in December 1979, following considerable turmoil arising from competition for political power in Kabul between different clients of the USSR. The Soviets took the Salang Pass, Kandahar and Kabul and killed one client president and replaced him with another. The war that ensued pitted the central government and its army, along with Soviet forces, against the rural, fundamentalist Mujahedin who rebelled against an irreligious government.

The wars of Central America have precedents reaching back to the 1920s. There was a small surge of revolutionary activity in the 1960s and a bigger resurgence in the late 1970s and 1980s, with the success of the Frente

Sandinista in Nicaragua in 1979, the failure of the Ejercito Guerrillero de los Pobres in Guatemala in 1982, and the standoff between the Frente Farabundo Marti de Liberacion Nacional (FMLN) and the government forces in El Salvador through the 1980s.

The long-standing dispute between Argentina and Britain over the Falklands (Malvinas) came to a head in April 1982 when the Argentinians overwhelmed the token garrison in Port Stanley. Overcoming their surprise, the British mounted their long-distance bid for recovery in the following month.

The war between Iran and Iraq that began in 1980 was the latest in a series between Persian and Arab, Shia and Sunni. It began when President Saddam Hussein demanded that Iraq be granted sovereignty over the Shatt al Arab in response to the Ayatollah Khomeini's calls for an Islamic, revolutionary overthrow of the secular, Baathist government of Iraq, and to a number of provocative border incidents.

Israel invaded Lebanon in June 1982 during Israel's fourth war with Syria in thirty-five years. The objective of the invasion was to forestall the threat of Syrian invasion and wipe out the presence of the Palestine Liberation Organization (PLO) in south Lebanon.

The Chinese army invaded North Vietnam in February 1979 on the pretext of defending the Chinese population in a disputed border region; however, this punishing expedition clearly sought to counter the Vietnamese invasion of Cambodia in 1978 and take the pressure off the Khmer Rouge.

The Somali invasion of the adjacent Ethiopian province of Ogaden in 1977 sought control of a region inhabited by Muslim Somalis where the rule of those in Addis Ababa had been disputed for some time.

When the Spanish relinquished control of the Western Sahara to Morocco and Mauritania in 1975, the Moroccans ignored the treaty stipulation that the locals' views be respected and annexed its part of the territory. This generated a rebellion by the Popular Front for the Liberation of Sakiet al Hamra and Rio de Oro (the Polisarios), which was supported by the governments of Algeria and Libya.

The Third World label, although not very informative geographically, did mark this sample of wars as typical of the political violence that has occurred since 1945. A geographical appraisal of these, then, provides a good starting point for a general study of warfare and the landscape.

GEOPOLITICS

Before examining the environmental settings and their strategic implications and then the significance of topographical detail for tactics, we must first establish the geopolitical contexts of these fights. This is a matter of the locations of the antagonists, of the disputed ground, and of the other sources of military and political power of consequence. Great differences obviously exist among these theaters of war; however, their locations far from the industrial core regions of the world imply long supply lines and limited infrastructure. The involvement of the great military and industrial powers in these disputes varied in general with the distance of the wars from their own borders. The USSR was directly involved in Afghanistan and the United States in Central America. Other than these backyard tussles, the wars were nationalistic squabbles fought where the hegemonies' presence was thin and local ambitions and animosities dominated the scene. The lack of immediate interest, the difficulties of projecting power and the possible cost of head-on collision far from home kept the two great powers in the wings in the political crush zone that lay between them.

Domination of Afghanistan was a symbolic rather than a strategic goal for the USSR. The military threat of an enemy's crossing the Amu Darya was slender, but the specter of encirclement and the fear of the Islamic revivalism spreading in central Asia were strong motives. Although Afghanistan contains the passes through the Hindu Kush to the Indo-Gangetic Plain, there is no conceivable temptation to command these as avenues for conquest. There were those in the American military establishment who claimed that the Soviet government had designs on Baluchistan as a path to the Straits of Hormuz where they could throttle Gulf oil traffic. However, the route through Afghanistan and Baluchistan is a long, dry, rough way to the mouth of the Gulf. Soviet involvement seems to have been a matter of propping up a client government reeling under the revolt of reactionary tribesmen. The Soviet position was not very different from the attempt of the U.S. government to bolster Reza Shah in Iran, except that the revolt there was too distant for effective direct intervention and too massive and urban to be forestalled by military action. The Soviets were left facing a rural revolt in a neighboring client state, a revolt aided and abetted by their distant rival.

The position of the United States in Central America was similar. The voiced motive of the U.S. military presence in the winter of 1989 was apprehension over political and social instability in what is deemed the American hemisphere. Strategic danger from Soviet military presence in the Americas was a very faint threat. The prospect of the Red Army invading from the south was fantastic. Soviet missile bases in the Caribbean would not have changed the nuclear balance dramatically. They would only have brought Soviet missiles within the same range as the U.S. weapons deployed in Europe pointed at Soviet targets. Khrushchev's effort to base missiles in Cuba was a desperate effort to outflank the U.S. early warning network when the superiority of Polaris and Minuteman became clear in 1962. By the 1980s this technological gap was no longer significant and there was no strategic benefit worth provoking the United States. The Panama Canal lost its military significance when aircraft carriers outgrew it. The Canal and the Caribbean seaways are not vital arteries to a still largely self-sufficient U.S. economy. Venezuelan oil and Jamaican and Guyanan bauxite are important, but not indispensable. What has replaced the Red Peril as the U.S. motive for interference is a selective concern for the moral and political tone of the continent. The mainstem of violence in Central America is plainly local. The U.S. government has committed its support to the landowning oligarchy, which puts it at odds with those who desire to reorder rights to property and power in Nicaragua, El Salvador, Guatemala and the rest.

The other guerrilla struggle, of the Polisarios against Morocco, lay outside the immediate reach and interest of either the United States or the Soviet Union, although both sides got equipment and weapons from the great powers. The phosphates of the region were not that enticing, although the United States clearly considered friendly command of the Gilbralter choke point important. The changing relationships among Morocco, Algeria, Libya, and the Polisarios complicated the picture, however.

The other wars in our sample are conventional clashes of nationalistic ambitions and hates, in which the United States and the Soviet Union had nothing to gain but could incur great potential costs and losses if they were to become engaged in them.

The seven-year struggle between Iran and Iraq in the lower valley of the Tigris and Euphrates was essentially drawn up along nationalistic lines, despite a strong undercurrent of Sunni and Shia contention and the Arab population in the Iranian oil-field province of Khuzistan. National loyalties remained largely intact in what was ostensibly a territorial dispute over the banks of the Shatt al Arab. The United States and the Soviet Union carefully kept the warring parties at arm's length, although the Americans favored Iraq through proxies, and the Soviets cultivated the sympathy of Iran.

Cutting through the confusion of overlapping claims of family, religion, faction and nation in Lebanon, the principal component of violence here was the battle for regional hegemony in the Levant between Israel and Syria.

The struggle for the Ogaden matched the irredentist ambitions of the Somali ruler against the centralizing grasp of Ethiopian imperialism. The switch of clients by the United States and the Soviet Union here was more suggestive of hegemonic gaming than the search for ideological compatibility.

China's punitive attack on Vietnam in 1979, playing out a long-standing animosity, had the appearance of a long-established regional power trying to cramp the style of a neighbor with aspirations to widen its radius of influence.

In the case of the Falklands it seems evident that the territory in question was not worth fighting for in economic or strategic terms. A coincident need for glory to divert attention from the unsatisfactory state of the economy and social fabric at home seems to have drawn the leaders of both Argentina and the United Kingdom into this affray. General Galtieri had to demonstrate competence, at least in his chosen profession. Unfortunately for him, for similar domestic reasons, Prime Minister Thatcher obviously felt the need to hang tough and demonstrate that the British imperial structure could still dish out punishment in distant waters.

THE THEATERS OF WAR

If geopolitics addresses the causes of war, we need to turn to physical and human geography to trace the course of war. The landscape presents targets for, or barriers to, violent ambition at the strategic or tactical level. When it comes to the fighting, topographic sheets or air photographs replace the broad locational picture of the 1:20,000,000 wall map of grand strategy as the relevant chart. For the soldier the landscape presents openings and obstacles for attack and defense. Hills, plains, rivers, forests, pastures, crops, roads and buildings become observation points, firing platforms, killing ground, areas of concealment and cover and avenues for movement. Along with the lie of the land and its artificial components, the weather and its changes affect the ease of movement, sighting and shooting. The winter snow of the continental interior transforms the fields of summer. Monsoonal flooding halts the operation of heavy armor.

The characteristics of the landscape that most concern soldiers are how far one can see and shoot and how easy the going is. The length of sight in an area depends on the ruggedness of the land, the vegetative cover, the weather and the disposition of hedges, crops and buildings. The ease of movement for men and vehicles is related to the array of rivers and canals, the slopes, the firmness of the ground, the density of vegetation, the types of field divides and size of fields, the variety of crops, the season and, most important, the availability of roads.

The most tactically significant aspects of the natural environment are relief and wetness. It is appropriate, then, to describe our theaters of war, working from the flattest and driest to the most rugged and dampest. This system reveals the strategic significance of these conditions.

Western Sahara

Rising up from the Atlantic coast to 1,000 feet, the western edge of the Saharan plateau is cut into a hilly terrain by the valleys of intermittent rivers (see Figure 2.1). The coastal plain gets nearly ten inches of rain a year between November and April, but rainfall diminishes with distance from the ocean. Mean monthly temperatures never drop below 50°F even on the coast in January, and in July and August the means are in the 80s.

The vast empty spaces of the Sahara present supply and movement difficulties for a mechanized army and enough cover and concealment to provide the defensive setting for guerrilla operations. It is difficult for regular

Figure 2.1
The Western Sahara

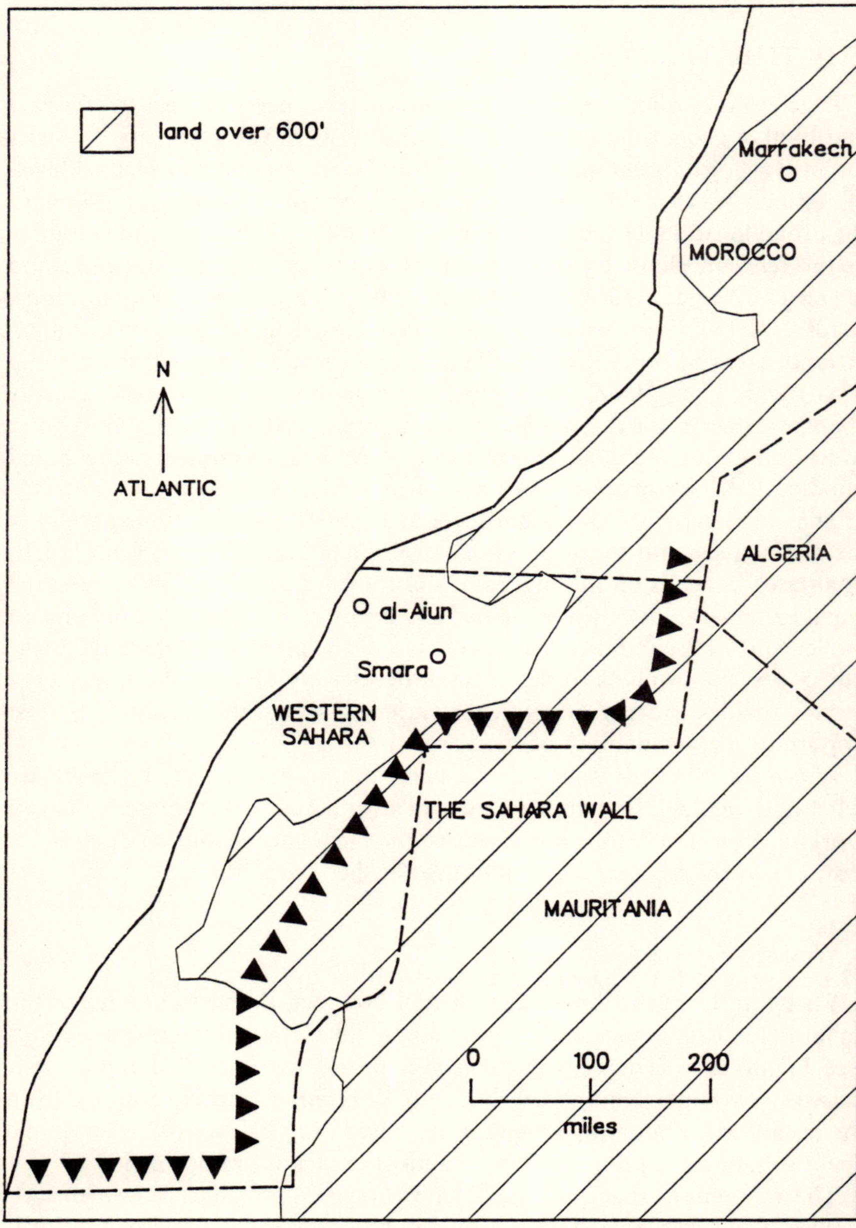

forces to nail down and destroy small, mobile bands of people familiar with the wadis who are practiced in stealth. There is little in the landscape to hold onto in the way of roads and villages from which to command space against insurgents. This also means, of course, that the offensive component of a guerrilla campaign is impaired by a setting with too few targets to hit in order to discomfort the government. The geography of activities was such that the Moroccans enclosed the vast majority of the population, the chief towns of al Aiun and Smara and the phosphate mines, which are the region's only exploited resource, with a 400-mile-long berm. This was subsequently extended into the 1,500-mile-long Sahara Wall, manned by 100,000 Moroccan soldiers. The approaches to this were mined and bugged. Forts and artillery emplacements were distributed along its length, and mobile reserves in the towns were prepared to reinforce any section that came under attack and to chase the attackers back to their Algerian bases. This fixed defensive perimeter stymied the Polisarios hit-and-run offense. Their response was to become more like a regular army by recruiting soldiers and by getting heavier weaponry, but by basing their strength on population numbers and international political clout, they gave up the advantages of guerrilla fluidity and thus, to some extent, their geographical advantage.

Iran-Iraq

The banks of the Shatt al Arab get about eight inches of rain a year with a rainy season, which lasts from November till April, that bogs down the off-road movement of wheeled or tracked vehicles during the winter. In addition the Tigris floods in spring, covering much of the region with snow melt water from the Anatolian highlands. The flat floodplain of the river is marshy and etched with abandoned river channels. To the north and east across the Iranian border the floodplain gives way to dry grasslands. Further north the border coincides with the foothills of the Zagros Mountains (see Figure 2.2). Straddling the Shatt al Arab there is a three-mile-wide ribbon of date gardens trellised with irrigation ditches. On the floodplain to the north of the confluence there is sporadic cultivation but the dry grasslands are used for grazing. The density of cultivation and population has never recovered from the Mongol invasion of the thirteenth century. Density of population is less than fifty per square mile over much of the area. The towns of the region are creatures of the oil industry.

In its initial attack on Iran, the Iraqi army had the advantage of short supply lines from its capital. Baghdad is only 75 miles from the closest point on the Iranian border and 300 miles from the Shatt al Arab. These miles traverse the flat plain of Mesopotamia. On the other hand, Teheran is 300 miles away from the border crossing leading to Baghdad and nearly 450 miles from Abadan, travelling over rugged terrain. Although this caused some supply problems, the distances did provide Iran with defensive depth,

Figure 2.2
Iran-Iraq

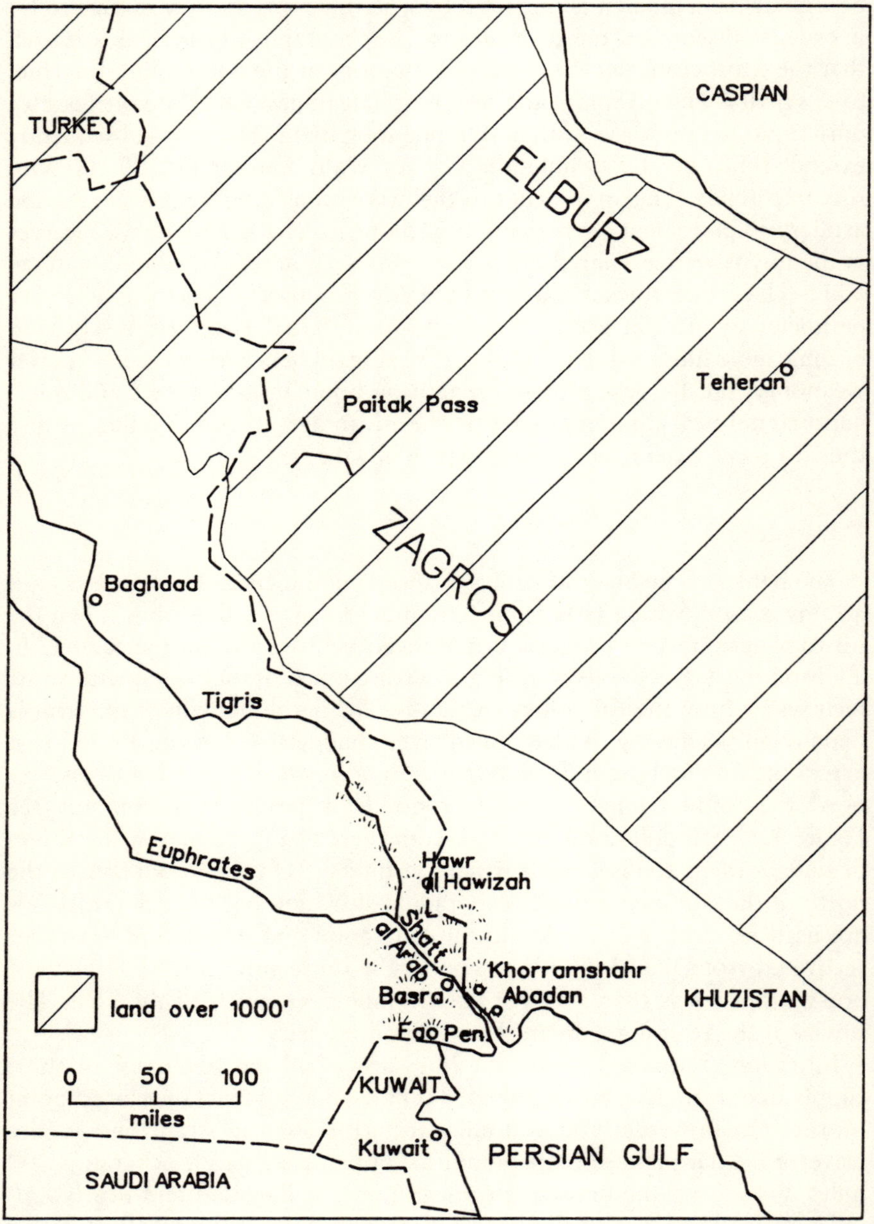

enhanced by the westward-facing escarpment of the Zagros Mountains. When the Iranians got the upper hand, the Iraqis were clearly at a geographical disadvantage as defenders. Basra and Baghdad had only a narrow territorial buffer, and to the east of Baghdad the opposition overlooked the defenses.

The Iraqi drive into the delta of the Shatt al Arab faced the disadvantage of the terrain for the attacker. Avenues of approach are confined by rivers, date gardens and swamps to the few roads with bottlenecks at bridges. The urban clutter of Khorramshahr and the string of villages on the road to Abadan slowed their advance. Their offensives were seasonally checked by rain and floods. In the dry season, heat and dust do not aid an army on the move. This is defensible terrain, and it is not surprising that the war ground down to an entrenched *sitzkrieg* with the occasional, if not seasonal, thrust and riposte.

Ogaden

The plateau of the Ogaden rises from the Somali border to the north and the west up the flanks of the Ahmar Mountains and the Audo Range (see Figure 2.3). Intermittent rivers running from northwest to southeast have scored valleys into this surface. Rainfall, from ten to thirty inches a year, is concentrated in the high-sun season. Drought-resistant acacia, which can survive in this setting, dots the savanna grasslands, except where the need for fuel has caused them to be cut back in a circle around the villages of the region. Throughout the year temperatures stay at around 80°F. Seminomadic herding and some farming in the valleys support up to ten persons per square mile.

As with Iraq's invasion of Iran, the Somali attack in the Ogaden struck at a peripheral region where the Ethiopians had defensive depth and the advantage of the broad lie of the land. Although the Ethiopians were faced with transport difficulties, given the distance from Addis Ababa, they could retreat to the high ground, trading territory for time. When the Ethiopians countered, the Somali's short supply lines were transformed into a perilously narrow defensive zone. Mogadishu and Berbera are only about 200 miles from the front, and the road between them, only about 20 miles from the border in places, could be cut by the Ethiopians at will. At the start of the war, the Somalis did enjoy the advantage of familiarity with arid conditions and the details of the terrain, inasmuch as many of their men had been fighting in the southern Ogaden as guerrillas since 1976. The Ethiopian army was in a strange and inhospitable setting. The Soviet equipment of the Somali forces was, however, not so well suited to the land.

Figure 2.3
Ogaden

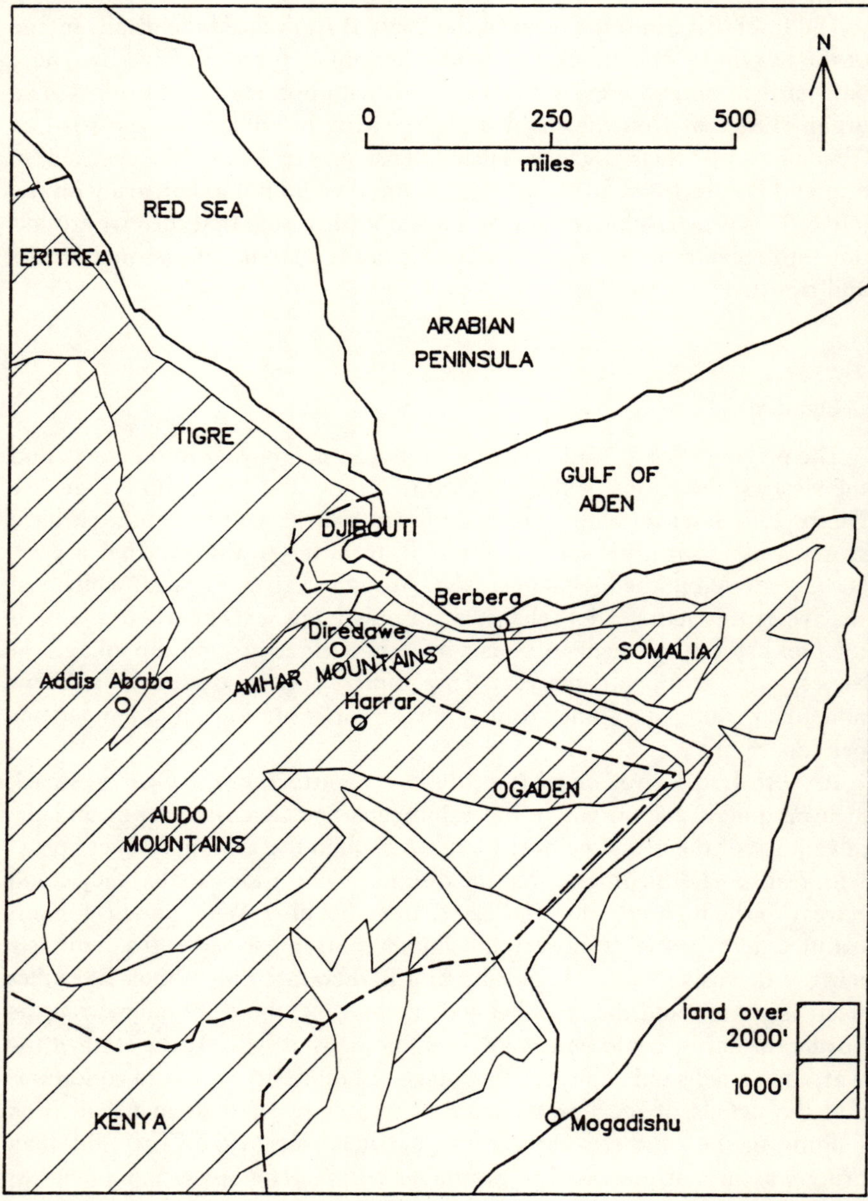

Lebanon

The grain of the Levantine shore of the Mediterranean runs strongly from north to south (see Figure 2.4). There is a narrow coastal plain interrupted by fingers of the Lebanon Mountains which reach toward the sea. The steep dissected slopes of the Lebanon Mountains rise to 10,000 feet above sea level. On their eastern flank there is a steep drop to the faulted trench of the Bekaa Valley which is occupied by the Litani River to the south. This five-to-ten-mile-wide valley with a relatively level surface is bordered in the east by the Anti-Lebanon Mountains which rise to a 10,000 foot crest. The climate of the region is Mediterranean with winter rainfall and summer drought. The coastlands get fifteen inches of rain a year, and the slopes of the Lebanon get as much as forty inches, leaving the eastern slopes, the Bekaa and the Anti-Lebanon in a dry rain shadow. The coastal plain is heavily cultivated; terraces carry cropping up the hillsides. The closeness of settlements and the density of population here are intense with cities occupying a large part of the land. The floor of the Bekaa is more sparsely settled. The mountains support a few pastoral people.

Apart from the obvious avenues of approach along the coast and the Bekaa, Sharon's plan called for a division-strong drive through the mountain backbone to secure the Dar el-Baidar col and cut the road from Damascus to Beirut. Despite their air and armor superiority, the Israeli *ugdah* was slowed appreciably in these confined battlefields. On the coastal plain citrus groves, villages, Palestinian refugee camps and cities blunted their drive to reach the forty-kilometer line within twenty-four hours. The central thrust by Einan's division through the Shouf Mountains was slowed by narrow, winding mountain roads so that it failed to reach its objective. The three-pronged attack in the east, up the center of the Bekaa and along its eastern and western foothills, turned into a full frontal slogging match. The two flanking drives were slowed by steep slopes and winding narrow roads, which presented a major difficulty in getting tankers up to fuel the spearhead. In the Bekaa Valley itself the Syrians fought fiercely for each road crossing, and the Israelis had no prospect of bypassing them.

Afghanistan

This land is high, dry and rough. The population is strung along the valleys which drain the Hindu Kush. Except for the area bordering the USSR along the Amu Darya River, most of the country is over 2,000 feet above sea level, rugged and rock strewn (see Figure 2.5). Temperatures in Kabul range from below freezing in January to around 80°F in July. Precipitation is equivalent to ten inches of rainfall a year in Kabul, fifteen inches in the mountains and three or four inches in the southern and eastern deserts. In winter this comes as snow, and Kabul gets a four-foot snowfall which melts

Figure 2.4
Lebanon

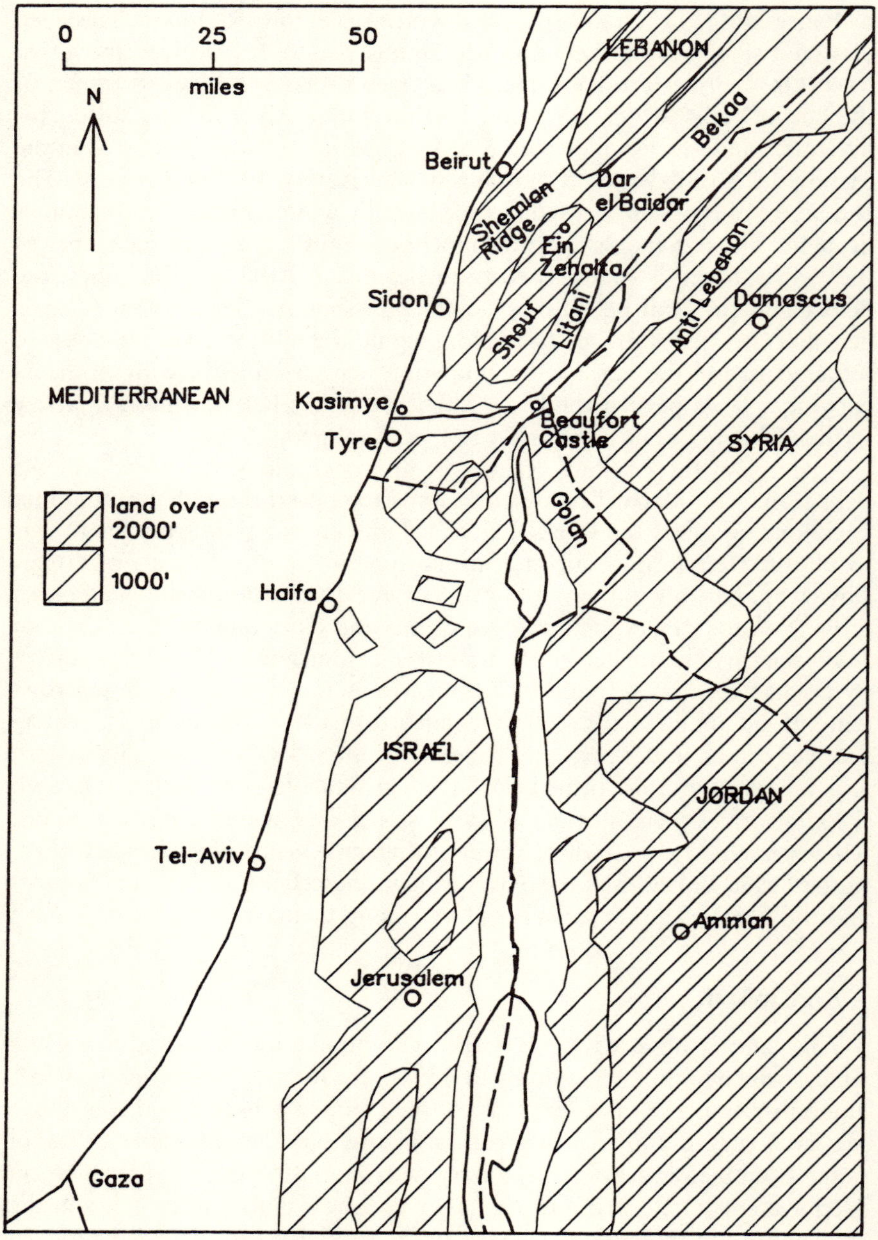

Figure 2.5
Afghanistan

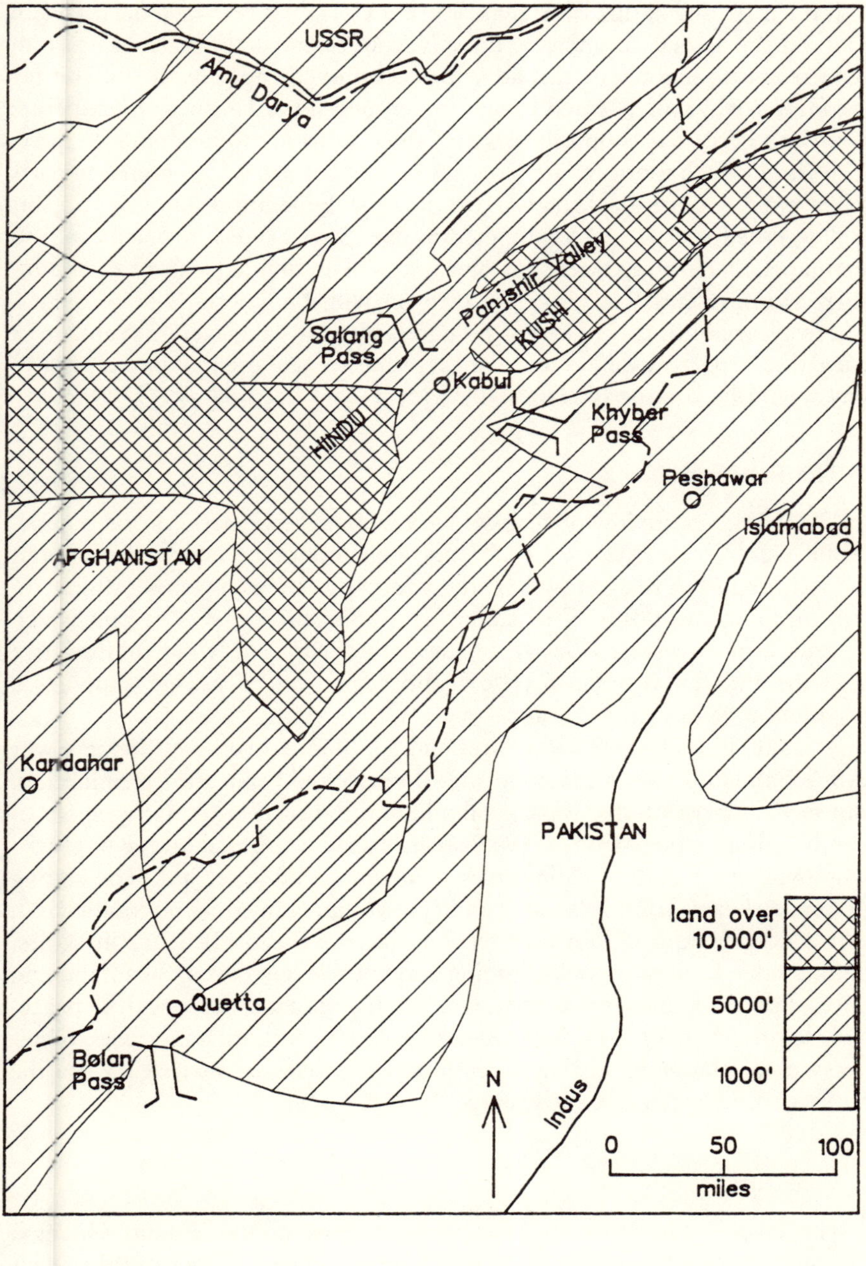

only at the end of March. In the moister north and east the vegetation is short grass steppe, fading into shrubby desert in the south and west. Above the valleys there are fewer than ten people per square mile; the highest reaches are empty.

If the fractured nature of Afghanistan's social fabric provided the occasion that led the Soviets to interfere directly, the fragmented nature of its landscape caused them to rue this adventure. It proved enormously difficult for a conventional army, geared to fight other mechanized armies on temperate plains, to carry out counterinsurgency operations in mountainous country. The Hindu Kush is not cavalry country. Soviet control was confined to parts of the cities and the main roads in the daytime. Expeditions into the Panjshir Valley failed to kill many Mujahedin, cost Soviet lives, and achieved no permanent territorial extension of control. The divided nature of the insurrection presented no target to turn on with hammer-and-anvil tactics. The way of life of the guerrillas prepared them to take every advantage of ruggedness, narrow defiles and cold weather to give the maximum grief to any would-be conqueror, especially with the aid of Stingers.

The Falklands

The Falklands are low plateaus of rock with jagged coastlines and bumpy, eroded surfaces (see Figure 2.6). The average monthly temperature is about 50°F through the year. However, the wind and damp lowers sensible temperature uncomfortably. The islands get about thirty inches of rain which drizzles from overcast skies throughout the year. The wind inhibits tree growth, and the vegetation is low and soggy. Sheep raising supports a population of less than 2,000 people.

At 1,000 miles from Buenos Aires and 8,000 miles from London, the war in the Falklands was a matter of naval reach and political determination. But beyond perceptions of national will and the successful management of combined operations from ships, British success was based on the practiced employment of mobile light forces in uncomfortable terrain. The setting precluded heavy armor. British training in yomping, moving on foot in cold, wet conditions through broken terrain, like the Brecon Beacons in Wales, paid dividends. Skill in rock climbing and knowledge of every headland and cove of the Falklands' coast, cultivated by individual officers at their peacetime leisure, were of great value in the operation. Although they were better dressed for the climate, the Argentinians showed no capacity to use the habitat to their defensive advantage.

Vietnam

The layered sandstone and limestone plateau of the Yunnan-Guangxi borderlands has been carved by rivers into a maze of steep-sided valleys

Figure 2.6
The Falklands

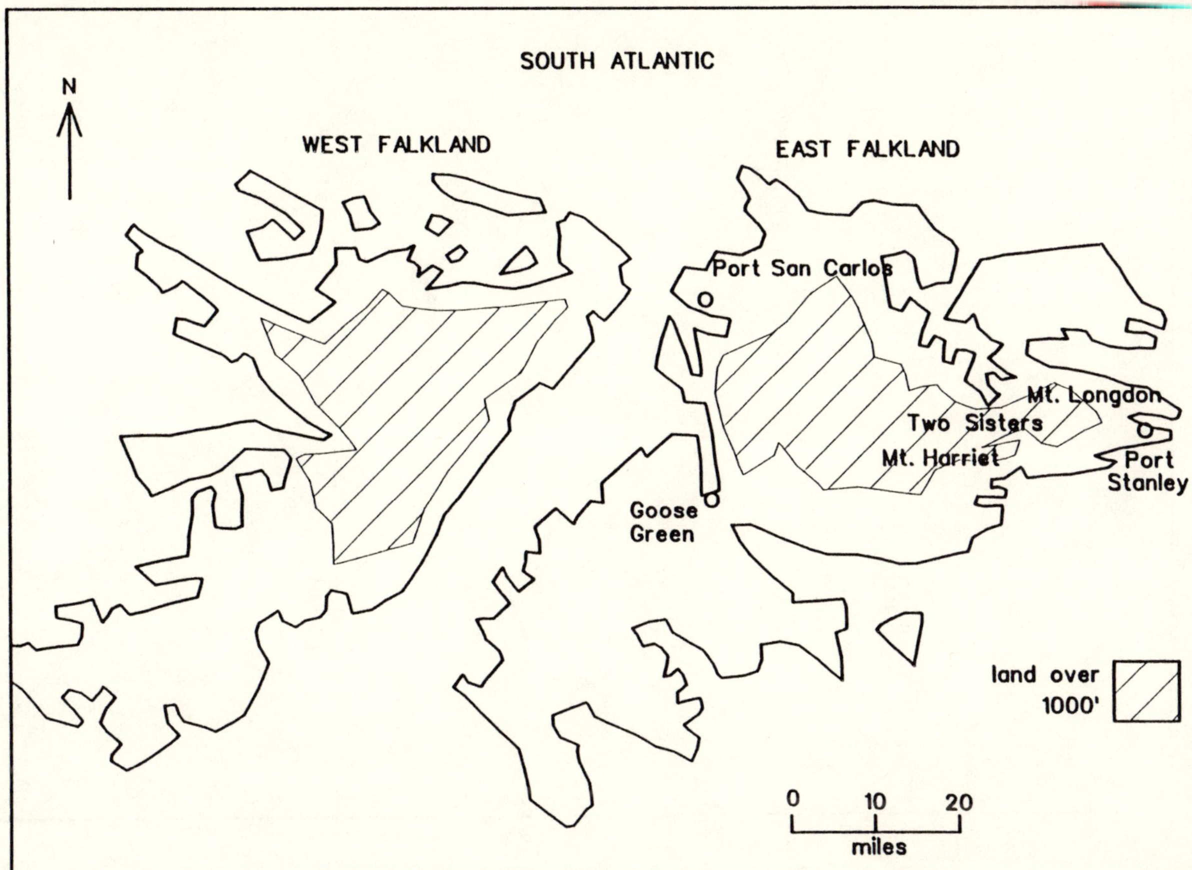

Figure 2.7
North Vietnam

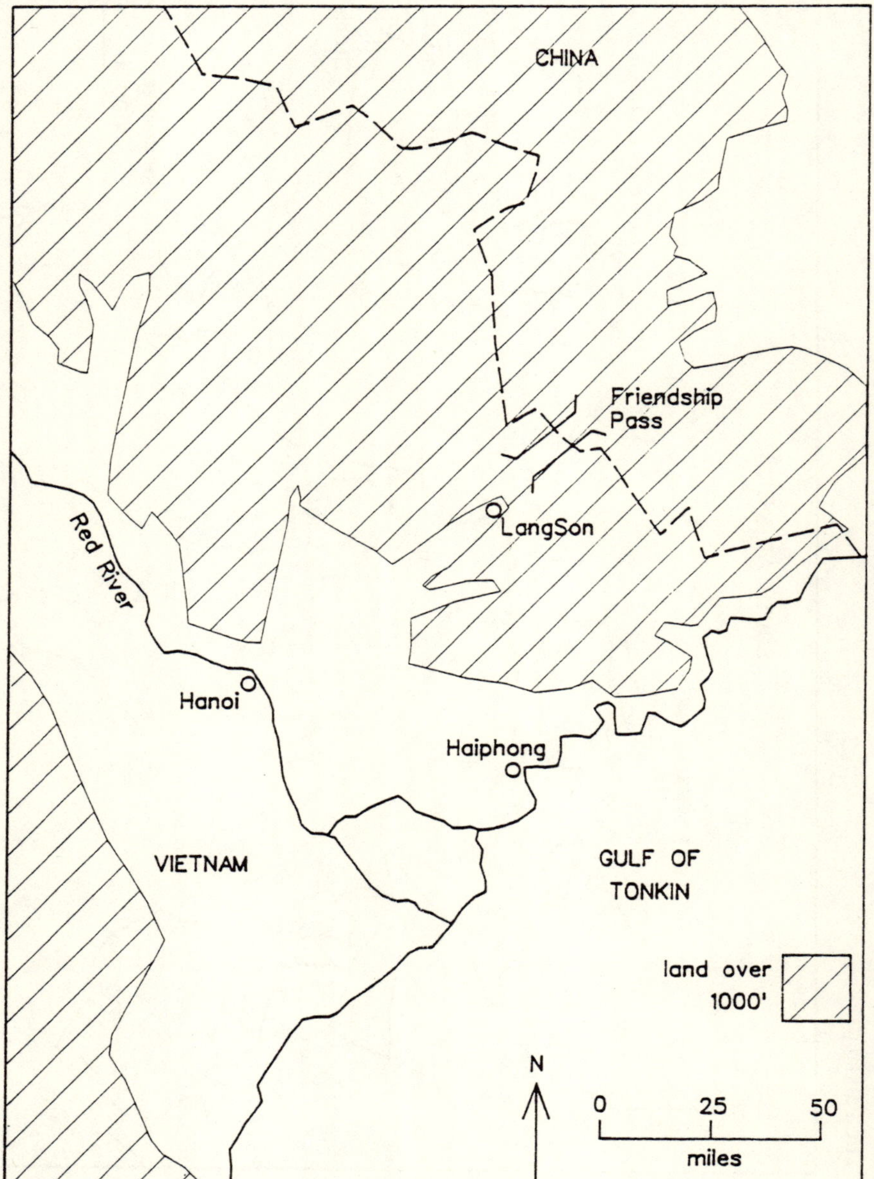

(see Figure 2.7). The climate of the region is monsoonal; over seventy inches of rain fall between May and September. The January average temperature is less than 60°F and July averages over 80°F. This rugged country is covered with deciduous forest with patches of savanna and cultivated clearings. Paddy terraces are cut into the gentler slopes. Density runs at about 100 persons per square mile.

The Chinese were very careful to confine their operations to the mountainous country of North Vietnam. They thrust as far as Lang Son, which guards the main pass from the north into the Red River delta, demonstrating that they could have attacked Hanoi if they had wished. Then they withdrew before the onset of the monsoon. In this symbolic gesture they did not wish to get entangled, to escalate their involvement and to pull the USSR into the conflict. The Chinese schedule took into account geographical conditions much further afield than North Vietnam. Winter freezing and spring thawing would keep the Soviets at bay along the Amur and Assuri border in Manchuria until May. Making their strike against Vietnam in mid-February gave the Chinese the greatest span of time before the onset of the monsoon would curtail their operations. Not only did they limit themselves in geography but also in weaponry. Little air support was used. Such weaponry might have invited a stepping up of the intensity of the conflict and would not have been very effective in this rugged, tree-covered landscape. Given the objective of the exercise and the nature of the land, Chinese operations were darting and mobile and made no effort to hold ground. Recognizing the nature of this campaign, the Vietnamese also limited their operations, discomforting the Chinese where they could with little cost and using the defensive advantages of the terrain in Fabian fashion until the monsoon could come to their rescue in May.

Central America

Except for the narrow coastal plains, valleys and high basins, Central America is ruggedly mountainous (see Figure 2.8). The region is wet and forested with over 40 inches of rain a year; over 100 inches of rain fall on the highest ridges. The wet coastal lowlands support tropical rain forest, giving way to low forest with patches of grassland on the drier northern margins. The middle slopes of the mountains are forested, as one climbs up, with broadleaves, then oaks, and finally pines on the upper slopes. The population is clustered in the small enclaves amenable to agriculture, with low densities of people in areas between. Even the clumps of population have low densities. The highlands of Guatemala and El Salvador and the basins of Costa Rica are the only areas with over 100 persons per square mile. There are few cities outside the capitals in this region.

Both government forces and insurgents spread out over the wooded mountains of Central America in the 1980s in their battle for the fealty of the

Figure 2.8
Central America

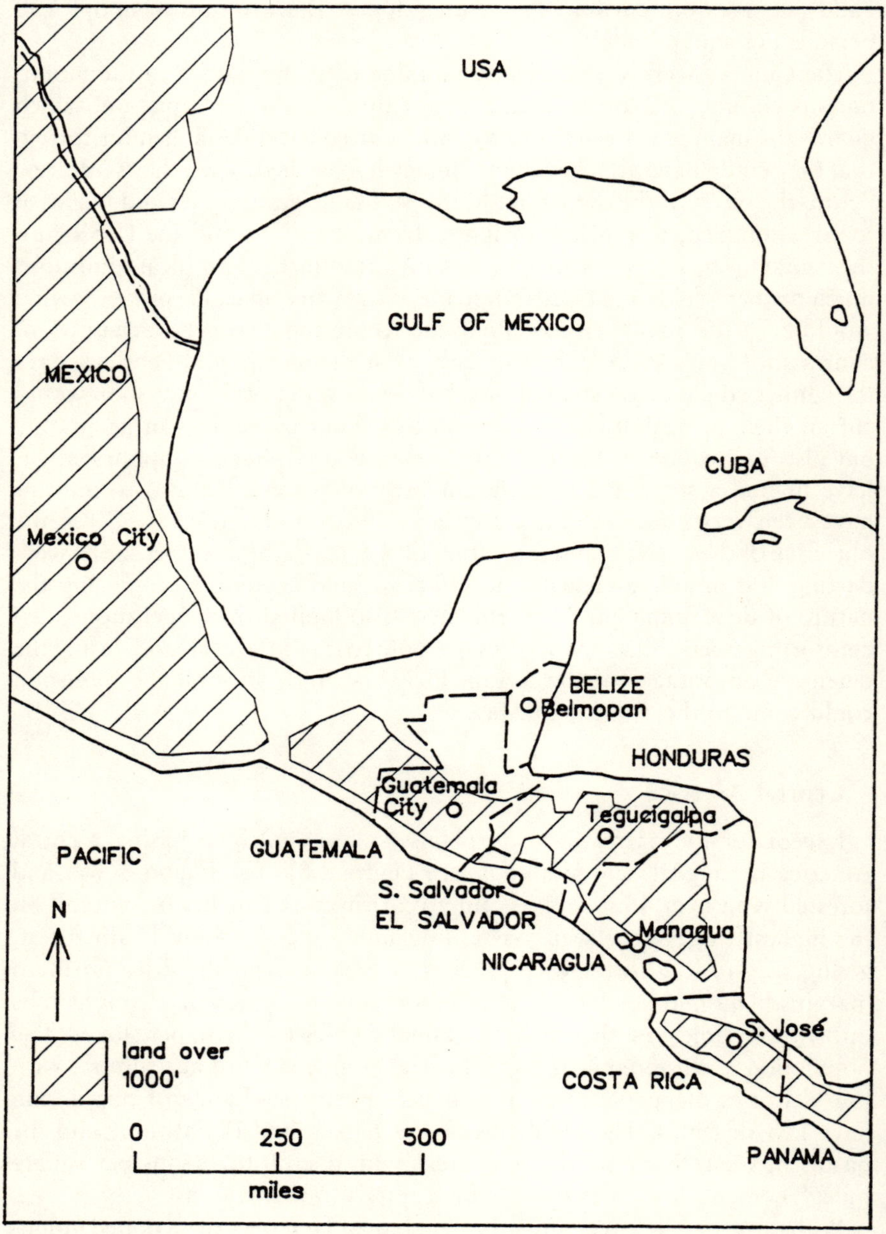

population. Che Guevara's revolutionary strategy of the *foco insurreccional*, manned by a staff of the wholly committed in a remote, rugged, unpopulated region, was abandoned for the doctrine of *guerra prolongada.* According to this second stratagem, the struggle for power is best spread out over time and space, bringing guerrilla war to peripheral but populated regions, drawing a greater part of the population directly and indirectly into the war. So, for example, in Guatemala the mestizo insurgents began to involve the Indians of the highlands in their efforts. This was countered with a new counterinsurgency ploy, especially by Rios-Montt in Guatemala. He substituted men for mobility. The flying column, the conventional method of destroying guerrillas, was replaced by tactical combat groups. More men in many small units were deployed through the contested areas of the highlands, establishing a more permanent military presence in the villages. This cut down the food, ammunition and manpower support for the guerrillas considerably. The Indians were armed and paid to kill guerrillas. From the insurrectory viewpoint there is no doubt that rough, wooded areas far from the big city provide a refuge from which to prolong the struggle. To be successful, however, guerrillas must operate an offensive defense. The government has to be at least irritated by constant and unpredictable attacks. In distant mountain areas there are few worthwhile targets. In the last resort, of course, in order to change the government, it is necessary to take the big city. The outcome of these conflicts in Central America, then, will depend on how well governments do at gaining and maintaining the confidence and support of the urban population or how well the guerrillas do when they come down from the hills and attack the cities. The guerrillas did well in El Salvador in the fall of 1989.

Environmental Conditions

These eight theaters of war bracket the variety of environmental conditions quite well. They range from flat, depositional plains and lower river valleys to mountains where erosion prevails, producing steepsided valleys and jagged interfluves. Climate extends from harsh aridity, through the monsoon, to the intertropical convergence zone, and even the cool damp of the Falklands. Combinations of elevation, global position and prevailing winds generate differences in temperatures and moisture regimes and, thus, in vegetation. There are vast tracts where it is too dry to sustain a continuous vegetative cover, where there is little to hide behind. As rainfall increases, the plant cover thickens from short to long grass, with a scatter of trees coalescing to clumps, then to the tangle of jungle and finally the high canopy of the rain forest. The human population of the rain forest is too slight to provide targets or occasions for big, bloody fights. Large numbers of men and machines cannot be moved and sustained easily in this setting. In general, the smaller the spaces between trees, the slower the progress of men and vehicles. A thick vegetative mass provides concealment and nearly point-blank sight lines. The jungle of the forest fringe and the savanna has seen more modern warfare. The seasonal shift of winds in this setting, the monsoon, introduces a seasonal element into military calculations. The rainy season brings floods to the lowlands, swells rivers and lakes, reduces visibility and makes action miserable.

There is a temptation to say that, in general, the landscape becomes tougher for conventional armies, the greater the relief and the greater the rainfall; however, very dry conditions are also inhospitable. The dust and heat of deserts take an environmental toll on mechanized warfare. What suits farming also suits war. The temperate plains, the croplands of the developed world, are probably the best ecological niche for warriors. One of the widely held misconceptions about geography and warfare is that deserts provide a sea-like arena for the likes of Rommel, Dayan or Sharon to deploy their fleets of tanks with nearly naval dash. This is not so. Once off the road, the going is tough in deserts. Sand desert provides poor traction and engine-clogging dust. Steep erosion surfaces on valley sides confine and channel movement. Bare rock pavement or the rough fragments of desert weathering do not provide good going. The boulder-strewn slopes of arid mountains keep tanks and trucks at bay and provide plenty of shelter and screening for ambushers. The desert can, then, be good defensive ground. The impression of speed and the fluidity of maneuvering in desert wars are outcomes of small numbers operating in a lot of space rather than an outcome of the desert's being a good medium for mechanized movement. The lack of vegetation, the clouds of dust raised by vehicles, the open fields of fire with long sight lines unhindered by trees or crops in desert plains allow long-range engagement and mutual destruction of tank armies at ranges of over a mile and a half. The lack of vegetation screening leads to attrition rates in desert warfare much greater than those experienced in temperate, wooded landscapes.

One common aspect of these eight war zones is a low density of human habitation and handiwork. In general, the population density varies with the availability of water and the steepness of slopes. Dry and rugged places are empty. Moister lowlands support a greater density. Rivers or ground water support strips or clumps of high-density irrigation farming. Apart from the Lebanon, the regions we have surveyed have sparse networks of roads connecting a scatter of villages. Towns are few and far between. Beirut, Tyre and Sidon stand out as exceptional scenes of city fighting in our set of conflicts.

TACTICS AND TERRAIN

When war comes down to the realities of killing and surviving, success often depends on a superior appreciation of the details of the landscape.

Before attacking or defending against an enemy in unfamiliar country, a shrewd soldier will look for places from which the surrounding country can be dominated. He will scout the approaches to these for himself and his opponent, noting any obstacles to movement. He will seek vantage points from which to see what is going on and select a killing ground where the enemy may be destroyed. Inversely he will find places where he can avoid being shot by hiding or by getting out of possible lines of fire. Even in the age of electronic sensing and control, of missiles, of fast-moving armor, air support and deadly first-round accuracy, the outcome of battle may still depend on whichever contestant uses the lie of the land to the best advantage. Time and again engagements, battles and wars have swung on the exploitation of some unique geographical feature. The successful adaptation to the general characteristics of a habitat with a particular tactic, weapon or formation depends finally on its use on a specific piece of the earth's surface, different in some aspect from any other. To act successfully in the face of the particularities of geography requires either luck or trained powers of observation and the ability to construct a mental image of the landscape. This latter talent, what von Clausewitz called *Ortsinn*, involves transforming information from verbal descriptions, topographic maps or air photos into a picture in the mind of the landscape, extending beyond the range of sight. Having done this it remains to predict the enemy's response to this setting and to act in such a way as to get the upper hand.

Experience and doctrine have produced a conventional scheme for terrain analysis in the form of a checklist of militarily significant components to look for in any piece of land. When synthesizing information from personal observation, maps, photos and descriptions, the soldier is advised to look for points of observation, fields of fire, cover and concealment, obstacles, key terrain, and avenues of approach. A more appropriate list of priorities, indicating the obvious sequence to use for such analysis, is as follows:

1. Key terrain
2. Avenues of approach
3. Obstacles
4. Points of observation
5. Fields of fire
6. Cover and concealment.

These six considerations become the headings to use for our exploration of these Third World wars at the tactical level.

Key Terrain

This term is used in a loose tactical sense to mark any bit of good cover, observation or firing platform, or a line of movement. There is, however, a higher, more strategic sense of the term, in which certain pieces of geography become the end of tactics rather than a means. The significance of some elements of the landscape extends far beyond the scope of an immediate engagement. There are crucial places that command the human occupation of the land over a wide ambit. Usually these are choke points on the road network, frequently at the limited number of points where a physical barrier is crossed. Time after time, the fight for a single structure or position focuses the unfolding events of a war.

Given the high cost and operational limitations of the helicopter, bridges over rivers can still provide the key to block or open the crucial link in any army's arterial network, which is why the Israelis raced north from Tyre for the bridge at Kasmiye over the Litani. Passes over mountain ridges have similar importance. The Mujahedin were able to demonstrate their disruptive power early in Afghanistan when they took the Salang Pass, cutting off Kabul from the USSR. The Paitak Pass, leading through the Zagros Mountains from Iran to Baghdad, was a key feature in the Iran-Iraq war. The Friendship Pass, commanding the route from Nanning to Hanoi, provided the Chinese with a geographical punctuation mark, with which to make clear the meaning of their demonstration of power to the Vietnamese. The Shemlan Ridge, overlooking both the coast road and the Damascus road, was a key position for the Syrians, providing a point of observation and a commanding field of fire over the approaches to Beirut.

In terrain where tanks are confined to roads, every road crossing becomes key terrain. The Israeli advance into the Bekaa Valley was measured by a series of battles for crossroads. In a lightly populated landscape, such as the Ogaden, road crossings and the only settlements coincide, and these were the scenes of the hottest fighting. When the Ethiopians managed to hold the principal keys of Harrar and Diredawe, they established the basis for their recovery against the Somalis. The fighting in the Chinese-Vietnamese borderlands was over the high ground that commanded the few routeways through this rugged land.

Avenues of Approach

To determine one's own best line of advance, or that of one's enemy, requires information on road and bridge capacities, slopes, loadbearing capabilities of the ground, and vegetation density. To these engineering aspects of route appraisal, however, we must add the need for the element of surprise. Surprising the opposition, and avoiding being surprised, can provide the necessary edge in an even match or enhance the strength of a weaker force. Surprise is partly a matter of timing but also a matter of selecting an unforeseen direction of approach. Geographical knowledge is needed to select the avenue that the enemy will least expect and to predict which route he will take. Intimate knowledge of the ways available has always been a major advantage of guerrillas fighting against an alien, regular army. At the start of the war in the Western Sahara, the Polisarios were most successful in using sheltered routes along wadis to hit Moroccan targets. Failure to guess the enemy's avenue of approach has accounted for many a defeat. The Argentinians did not prepare for the most obvious British approach to Port Stanley. The British landing at San Carlos, the only possible site for such an operation on East Falkland, was virtually unopposed.

The availability of avenues of approach may be a matter of sheer capacity. Chinese superiority in men and tanks was severely diminished by the lack of road capacity in the broken, wooded terrain of their border with Vietnam.

Obstacles

Over much of the world's landscapes, where slopes are steep, the woods thick or the going soft, mechanized armies are virtually confined to the road. The rest of the terrain is one enormous obstacle from a military viewpoint. There are degrees to this, and for a while the helicopter provided the prospect of overcoming surface obstacles in a tactic of three-dimensional envelopment. The Stinger and its descendants may inhibit this, and obstacles do still present a defensive advantage. The extreme case of forbidding terrain is found in Afghanistan, but, even on the plains, natural and manmade obstacles can channel and halt an aggressor's movement. The Iraqi drive in the spring of 1981 to encircle and capture Abadan was blocked by the seasonally flooded land to the east of the Shatt al Arab, where the date groves lining the river provide a dense screen protecting the road southeast from Abadan to the Gulf. On the western flank of the Israeli advance into Lebanon, the towns and Palestinian refugee camps of the coastal plain presented a series of artificial obstacles to an armored advance.

Some of the greatest feats of arms have involved conquering physical barriers and using difficult terrain as an avenue of approach to surprise the enemy or overcome his advantage in position, numbers or equipment. Hannibal crossed the Alps. Wolfe scaled the Heights of Abraham. The Ardennes were used by von Rundstedt and von Manteuffel in 1940 and 1944. Success is not guaranteed and persistence helps. The Iranian March offensives of 1984 and 1985 struck toward the Baghdad to Basra road through the marshes of the Hawr al Hawizah at a time of year when flooding would offset the Iraqi superiority in armor. The Iraqis halted these drives with great loss of life by dint of their greater firepower and air cover. Then in February 1988 the Iranian thrust finally overwhelmed the Iraqi firepower with numbers and captured the Fao Peninsula.

Points of Observation

Despite the artificial aids for sensing the enemy that are available now, it still pays to keep a naked eye on what is going on. There is a big advantage to seeing your enemy before he sees you. The Mujahedin of the Panjshir Valley were in no doubt about this when they climbed into the mountains to watch the Soviets thrash about below. The Argentinians put themselves at a big disadvantage when they did not man the high ground around Port Stanley properly and conceded Mount Longdon, the Two Sisters and Mount Harriet to the British, giving the British the advantage of sight, intelligence and initiative.

Bitter fights still take place for a vantage point. The Israelis threw themselves into dislodging the PLO from Beaufort Castle overlooking the Litani Valley. In this case, however, the significance of the site was as much historic and symbolic as it was operational. The Syrians and the Israelis fought fiercely among the terraced fields on the flanks of the Shemlan Ridge to see who would overlook the coastal approaches to Beirut from the south as well as the road to Damascus.

Even in war in the air with electronic sensors, whether or not one can see the enemy and the ground one chooses to keep a lookout from still matter. The Israeli air force came at the Syrian defenses in the Bekaa out of the sun. The Syrians had to turn to radar to sense them. Unlike the human eye, however, radar did not distinguish between drones and Phantoms, thus, many missiles were spent on dummy targets. Furthermore, switching on radar made the missile sites vulnerable to antiradiation weapons.

Fields of Fire

One eagerly sought advantage terrain can afford is a killing ground where the enemy can be shot with little chance to reply. One of the objects of terrain analysis, then, is to seek firing points that cover an avenue or a field that the enemy cannot avoid.

The Israeli central axis of advance through the Shouf Mountains was stymied at Ein Zehalta. The Syrians brought them to a halt at this juncture two miles short of their goal, the Beirut-Damascus road. This had wider ramifications inasmuch as it lead Sharon to press for the air strike at Syrian missile sites in the Bekaa Valley. The Syrians held the Israelis at Ein Zehalta by establishing a superior field of fire. Just to the north of the village the road winds down into a wadi. A Syrian brigade stationed its tanks on the northern edge of the wadi and its infantry on the wadi floor, covering everything that arrived from the south hoping to cross the valley. The point of the Israeli column was hit when it arrived, and it had little to no chance of returning fire successfully.

Even guided missiles are limited in their effectiveness by inhibitions in their field of fire. To direct them to their targets it is necessary to get locational information. If the landscape throws shadows across the sensory field from which these data are derived, then there is cover for an approaching enemy to exploit. The steep sides of the Bekaa Valley masked the coverage of Syrian radar stations located in the valley floor. Israeli aircraft coming in over the mountains from the west were, thus, sheltered from Syrian ground-to-air missile batteries until the last minute. The Israelis were, therefore, able to destroy these batteries with little risk in 1982.

Cover and Concealment

Guerrilla fighting is to a large extent an exercise in the exploitation of cover and concealment. Ambush and the sudden hit from a stealthy approach are the chief guerrilla ploys. The lightly populated, wooded uplands of Central America and the jagged, dissected mountainsides of Afghanistan provide a lot of hiding places. At the other end of the population density spectrum, the narrow passages and irregular structures of refugee camps provided the PLO with their only effective defenses against the Israeli army. Even if reduced to rubble, they still provided protection and concealment. The warren of shelters and bunkers in the camps on the edges of Tyre and Sidon stalled the Israeli advance significantly. The PLO held out for a week in Ein Hilweh on the outskirts of Sidon. Even a farmed rural landscape can provide cover. The orange groves of Lebanon were used by the PLO as cover from which to ambush tanks. The date groves of the Shatt al Arab not only were an obstacle for tanks to get through but also provided screens to hide movement from enemy eyes.

Weather still provides screening for conventional operations. The British landing on East Falkland was delayed until fog provided cover for the approach of the assault force to Port San Carlos.

The shelter and concealment offered by some landscapes can cancel the effectiveness of particular weapons and leave the outcome to the balance of other arms. The tree-covered hills of northern Vietnam provided such good protection against air cover that neither China nor Vietnam used air support for attack or defense. This suited the Vietnamese in terms of their supply of arms and the Chinese from the political angle, since they wished to fight a confined and limited war.

CONCLUSION

This examination of some recent wars suggests a sequence of matters that need treatment in a study of the geography of warfare. Although the chapter began with a discussion of geopolitics, it seems more appropriate to set the physical and human geographical scene first to provide the matrix for a subsequent geostrategic analysis. This examination of geopolitical issues calls for a specific consideration of revolutionary war. The kernal of the book lies in a treatment of tactics and terrain that involves a shift down from the strategic to the operational level. In particular we will focus on low-intensity warfare and urban warfare. As a sustained example, Chapter

11 considers the case of the British Army and the IRA in Northern Ireland. The last chapter turns to the motives of those who fight and the part that identity with place and, especially, nationalism play in bringing people to kill and be killed.

REFERENCES

Harkavy, R., and S. G. Neuman, eds. 1985. *The Lessons of Recent Wars in the Third World*. Vol. 1, *Approaches and Case Studies*. Lexington, MA.: D. C. Heath.

Neuman, S. G., and R. Harkavy, eds. 1987. *The Lessons of Recent Wars in the Third World*. Vol. 2, *Comparative Dimensions*. Lexington, MA.: D. C. Heath.

3

THE LIE OF THE LAND

The last chapter surveyed the geographical circumstances of eight recent wars. In this chapter the survey is extended to the global scope in the form of a commentary with which to peruse the thematic section of a world atlas, such as *Goode's World Atlas*, building up the picture of human habitation on top of the physical basis. Since a map is worth a thousand words, clearly, what is offered here is not a substitute for looking at the maps. Rather this is intended as an encouragement and an aid to digest the contents of an atlas.

GEOSTRATEGY AND GEOGRAPHY

The first purpose of this exercise is to set the scene for a discourse on the geographical incidence of war. War is a geographical phenomenon. Its occurrence is not random. There is a distinct geographical pattern to war. Geopolitical theorizing has, in effect, conjectured on the location of wars inasmuch as it points out where imperial and national interests will compete for command of territory. The variables that enter into the explanation for war include global position, resources of land and minerals, population, transportation and military technology. A macrohistorical viewpoint of human conflict, then, requires a macrogeographical frame of reference. A description of the main patterns of physical and human geography is a proper preface to any discussion of geostrategic issues.

TERRAIN AND TACTICS

The second purpose of this chapter is to provide some detail on the chief components of the world's variety of landscapes which come into play in deciding the outcome of war. These are the elements of our various habitats which come into focus when we scale our vision of the world up to the finer level of resolution of tactics and military operations. Here we will explore some of the features of the range of the earth's environmental settings that accrue to the advantage either of the defender or of the attacker in the military balance.

Fighting is mostly a matter of moving, hiding and shooting. The ability to do all three is conditioned by the lie of the land and climate. Apart from questions of numbers, weapons, vehicles and organization, the outcome of war often depends on who best employs the advantages offered by the physical and human elements of the landscape for defense or attack. In more traditional contests for power, where the object was subjugation rather than slaughter, the parties to the dispute found a suitable open patch and set about their rites of violence. When the action gets serious and annihilation or survival is the goal, then ambush, the stealthy approach and holding the high ground are age-old ways of using geographical intelligence to increase the effectiveness of men and weapons. Some groups gained a martial reputation because their rugged homeland gave them a defensive advantage. Meanwhile, dwellers on the plains had little choice but to submit, flee or die in the face of a better armed conqueror. The cultural diversity found in uplands like the Alps, Balkans, Caucasus and Himalayas reflects the defensive advantage. The large areas of homogeneous culture in lowlands suggests the advantage of the aggressor in this setting. It might be countered that mountains provided a refuge because they were not worth conquering. There are, however, enough instances such as Thermopylae, Marathon and Obwalden to support the thesis of the defensive advantage of an upland home base.

In general, if movement is easy and if there is not much in the way of cover and concealment in a certain habitat, the advantage lies with better equipped aggressors. If the going is rough and if hiding places are abundant, a poorly armed defender can hold off a would be conqueror more easily.

Where greed, ambition for power and pride bring peoples and nations into violent conflict, the outcome, as well as the prospects for a peaceful solution, hangs to a considerable extent on the climate, soil, vegetation, relief, drainage, and disposition of roads, fields and buildings in the territory in dispute. Competition for land, resources and political power has escalated into the violence of war in a variety of landscapes. The combination of climate and surface material of the land provides a variety of soil and vegetation and, thus, of human habitats. In many places the ground cover and the details of the landscape have been greatly changed by hunters, farmers, miners and town dwellers.

To survey the settings in which war may take place, we need to consider not only physical geography but also the human response and its marks on the land. We will consider first the global variety of physical settings and then examine the disposition of people and their handiwork.

CLIMATIC VARIATION AND HUMAN RESPONSE

To trace the pattern we begin at the equator and work toward the poles.

The continental tracts and islands that are strung along the equator are covered in a belt of hot, wet rain forest. Here temperatures never fall below 65°F and are usually in the 80s. There is little seasonality of temperature or rainfall. This comes in torrential downfalls amounting to more than eighty inches a year. Beneath a surface layer of rotting leaves, the soils are heavily leached. The trees are tall, broad-leaved, heterogeneous evergreens with a dense canopy more than a hundred feet above the ground. This shades out much in the way of undergrowth, but there is a lot of hanging growth, such as lianas. Although it is comparatively easy to move through on foot, it is hard to navigate and maintain contact here. Vehicles are confined to the rivers and the very few roads. Access from the rivers is impeded by dense galleries of mangrove and lower growth. Roads are ripped into gullies by rain shortly after they are built. Where the forest has been hacked and burned by shifting cultivators, loggers or graziers, the second growth is jungle. Very dense at the ground, this consists of more homogeneous stands of smaller trees and brush, which is very difficult to penetrate. Where the land rises up above coastal plains or river basins, the difficulties of the vegetation are compounded by very rugged terrain, caused by the deep dissection of heavy runoff.

The rain forest is the last refuge of a handful of hunters, and it houses a scattering of from two to twenty-five shifting cultivators per square mile. There are some patches of higher population density and a meager network of roads and settlements. The greatest densities in the equatorial world are found where soils formed in mineral-rich, volcanic deposits offset the leaching of heavy rainfall and allow for the multiple cropping of rice, as, for example, in Java and the central Philippines.

A seasonality of rainfall sets in with distance from the equator. The effect of continental heating and cooling causes a reversal of winds blowing across the continental margins. When the sun is overhead, the landmass is heated and low pressure draws moisture-laden air onshore for a rainy season, which diminishes in intensity and length with distance from the equator and the coast. When continents cool, an outflow of air from the high-pressure system formed brings a dry season. As the length of the dry season and dryness increases, rain forest gives way to thicker, lower, partly deciduous forest and bamboo. This cover survives in the uplands. In Asia the lowlands in this setting have been cleared for growing paddy rice. The dry season provides a time for ripening and harvest. When rainfall is less than forty inches a year, wheat is the staple. These lands support enormous numbers in closely spaced villages linked by the low-level transport network of a peasant economy. Towns, spaced about ten miles apart, draw on the limited surplus of the countryside. At focal points cities with millions of people dominate the human landscape administratively and commercially. These are surrounded by vast shantytowns, which house refugees from rural poverty. Despite this presence, these are still primarily farming societies.

Where it is drier, the forest gives way to savanna with low, widely spaced trees, shrubs and tall grass. This is a vegetational transition between the forest and the dry grasslands of the continental heartland, which cover the interior of southeast Asia and India. Elsewhere it extends along the northern fringe of Australia, and in a great arc from Gambia through the interior of west Africa, across the eastern plateau, sweeping around the edge of the Congo rain forest, reaching south to the Zambezi and along the coast of Mozambique to the Limpopo. Across the Atlantic, the southern portion of the Amazon basin, the Brazilian highlands, the Orinoco basin and much of central America are savanna. Its leached, red soils will support only the light density of shifting cultivators or herdsmen. These activities, traditional or commercial, generate little infrastructure. In the dry season movement is easy, but things tend to bog down and rivers flood in the rainy season.

Where rainfall is less than twenty inches a year, grass shortens and few trees and shrubs are strung along the water courses. Such a landscape fringes the deserts of Australia, southern Asia, the Sahara, the Kalahari and the Namib. It accounts for most of northern Mexico, west Texas and New Mexico. This is the domain of herders, either tribal nomads or commercial ranchers. In this dry setting with only occasional, heavy runoff, the banks of rivers and valley sides can be fairly steep. Slopes are sharp. For the most part, however, this is easy terrain to travel, with little cover.

With less than ten inches of rain a year, the vegetation cover is discontinuous in a desert landscape. With a light sprinkling of drought-resistant grasses, cactus or sagebrush, much of the desert lacks a true soil. Most of it is bare or rock-littered pavement, which makes the going very difficult on foot or in a vehicle. The depositional declivities of desert lands are filled with sand which blows into a flat sheet or riffles in a variety of dune forms. This is not easy going, especially since sand plays havoc with humans and machines. The outlines of the desert are stark, with steep slopes marking any change of level, footed by an apron of unconsolidated detritus. The traditional routes of desert crossers are the wadis, the dry valleys of intermittent or ancient streams. Along these wadis, or where rivers run down from moister mountains, small oasis settlements punctuate the limited feasible routes for desert movement. The dry valleys do present a slight risk since they may carry a flash flood when highly localized, heavy rain runs rapidly through a landscape with little to absorb and reduce its headlong rush. Dry mountain landscapes, such as the Hindu Kush, are rugged, barren and rock strewn. In some climatic circumstances mountains in desert areas may pull down enough rainfall to support some greenery. These uplands, like the Atlas, the Hijaz and the Zagros, sustain a modest density of farmers and graziers. Their rugged terrain and cover have provided protection against invaders through the centuries for Berbers and Kurds, for example. The significant rural population and the biggest cities of the desert realm are found in the linear oases where the great rivers—the Nile, the Tigris and Euphrates and the Indus—bring tropical runoff or mountain snowmelt to flood the plains. Strips of multiple cropped, irrigated fields, date groves, compact villages and great cities are confined to the flood plain. Beyond lies desert emptiness. The oil industry has added its creatures to this setting in the Gulf states and north Africa in the second half of this century, along with concrete political mirages like Riyadh, sustained by oil revenues. The extraction of metals, nitrates and phosphates elsewhere brings small temporary towns and facilities to move large quantities of material to desert landscapes, for example, in northern Chile, the American cordillera and western Australia.

On the western margins of the continents between 30° and 40° of latitude, the desert becomes seasonal. Only in the high sun season does stable, dry air prevail. In the winter months the belt of westerlies shifts equatorward and brings onshore winds and rain to the shores of the Mediterranean, to California, central Chile, Cape Province and the southern tips of Australia. The evergreen forest, scrub and parkland, remnant in the maquis or chaparral of the uplands, was replaced by vineyards, olive groves and wheat fields. In the Mediterranean basin itself the human landscape is attuned to war, with big villages and fortified towns commanding key terrain. Groves of citrus and other tree crops and irrigated fields of vegetables have taken over much of this landscape in this century.

Between 40° and 70° of latitude on the western littoral, the westerlies prevail through the year, bringing moist, mild conditions and deciduous forest in latitudes where the seasonal variation of the angle of the sun is more marked. These maritime influences penetrate deep into Europe where the mountain ranges in general trend east-west. The mountain spine of the Americas confines this climate to the Pacific northwest and southern Chile. Elsewhere it touches Tasmania and New Zealand. In Europe it sustains intense crop and animal raising, with a stronger pastoral component in the wetter uplands and more remote regions. The fairly constant and steady runoff generated by this climate and its vegetative sponge, produce gentle, rounded slopes and a rolling terrain. This setting in Europe saw the beginnings of industrial and urban growth from 1750, and, where coal was accessible, along the border between the European plain and the central, Hercynian uplands, great swathes of row housing focused on mines and factories developed, commanded by metropolitan clusters of residence, commerce, administration and conspicuous consumption.

On the eastern margins of the continents between 20° and 40° of latitude, the easterly trades bring rain through the year to a warmer setting. The humid subtropics have leached soils beneath their forest cover in eastern Asia, southeast Australia, Natal, southern Brazil, and the American South.

In China, Japan and Korea, the lowlands with this climate are intensely farmed, with terracing carrying cultivation and paddy fields up the hillsides. Industrialization has been accompanied by the growth of densely packed cities in the small coastal plains of Japan and Korea and the great river valleys of China. Elsewhere European settlement intruded into these regions in the nineteenth century, cutting back the forest to grow sugar, cotton or coffee. Since the passing of this plantation era, a more stable, diverse agricultural economy has evolved in a landscape of big farms and small towns with a sparse population amidst woodland, pasture and crops. Industrial and urban growth has also arrived in selected parts of this environmental niche during the last century, for example, along the piedmont and coast of the American South and in the hinterland of São Paulo and Rio de Janeiro.

In the interior of the continent in the midlatitudes, on the cooler margins of the deserts, limited rainfall becomes more effective because of lower evaporation. Discontinuous vegetation thickens to the short grass cover of temperate steppe. Where rainfall is over fifteen inches, the chemically neutral, organic richness, which had developed beneath the sod and was grazed for ages by wild and domesticated herds, was turned up with steel ploughs after 1850. These grasslands became the breadbaskets of the world. The greatest belt reaches from Kiev to Omsk. Its north American counterpart stretches south from the Saskatchawan River to the Panhandle of Texas. From Rosario a golden band crosses the pampas to Bahia Blanca. This same economy dominates the hinterlands of Perth, Adelaide and Melbourne in Australia. This is a landscape of vast wheat fields on farm units of a square mile or more with buildings a mile or more apart. Small towns thirty miles apart on railroads or waterways, built around grain elevators, organize the landscape. In Eurasia this was the home of nomadic light cavalry. There was little to hinder movement or provide cover in summer. The snows of winter, however, make this a very different proposition for movement and fighting. The horseman has been squeezed out between the fluctuating margins of the desert and the extension of the frontier of farming with steel ploughs, water pumps and improved breeds of cereal. In the long run the horsemen of the steppes could never hope to win against the greater density of the cultivators and their capacity for producing weapons.

Where rainfall exceeds twenty inches a year in the middle latitudes, farming intensifies to produce a combination of crops and livestock in humid, continental climes. Here the prairies give way to deciduous woodland in the heart of north America and on the Russian plain. The forests have been cleared to sustain an economy of compact villages or individual farmsteads, housing over sixty persons per square mile in areas of richer soils. A tight network of service centers is strung together by a dense network of roads and radial spokes of railroad. The urban industrial economy has spilled over into the most accessible of the locations in this setting since 1800, and these areas now house 90 percent of its population. Merging with the maritime regions of Europe, this was the scene of the closing stages of World War II and the expected theater of confrontation between the might of the United States and the Soviet Union up to 1990. In the continental interior and east in the midlatitudes, seasonality becomes strongly marked. Summers around 70°F with thunderstorms bring four months of hot humidity. Winters in the 30s bring snowfall. Fall and spring lie between these extremes. As heating thins out with the more oblique angle of the sun in the northern tier of America and Eurasia, grass and deciduous forest give way to coniferous trees in the taiga which stretches from Alaska to Labrador and from Norway to Kamchatka. Beyond the Arctic Circle, where the warmest mean monthly temperature falls below 50°F, we reach the treeline and the tundra beyond. With a permanently frozen subsoil, this turns into a quagmire amidst the mosses, lichens and sedges in the brief summer thaw of the surface. Beyond this lies the implacable cold of the Polar icecap.

These extreme conditions are found elsewhere when the land rears up and the temperature drops by 3.6°F per thousand feet. In mountain country, even at the equator, there is a horizontal zoning, ringing the changes of climate and vegetation through the whole gamut to the tundra of alpine meadow or, indeed, permanent snowcap. The steep slopes of uplands, where erosion exceeds deposition, do not allow for much in the way of soil development. In the tropics, however, elevated coolness in high plateaus and basins is very desirable for human habitation. This accounts for the attraction to east Africa, central America and the Andes of a modestly high rural density.

THE HUMAN PRESENCE

The geographical distribution of people and their handiwork is highly variegated. There are great tracts of land with very little human habitation. Two-thirds of the surface of the land is inhospitable because it is too dry, too wet, too cold or too rugged for farming.

The dry limit in the midlatitudes is the fifteen inches of rain needed to support nonirrigated agriculture. With higher evaporation in the tropics, the limit is thirty inches. The dry grasslands of the world support a few people who raise livestock. Exotic rivers and oases create strips or dots of rural density, and minerals attract urban settlements to the deserts.

By contrast the equatorial girdle is too wet for farming. The fragile soil of the rain forest cannot support shallow-rooted food plants for more than a few years before its heart is depleted. Shifting cultivation or gathering sustains a few. Exceptions arise where rich volcanic soils offset leaching, or where commercial tree crops, such as rubber, bananas, cocoa or oil palm, sustain fair densities in Malaysia, west Africa and central America.

The most serious limit to farming is cold. Most food plants need ninety frost-free days to mature. This excludes a vast acreage of North America and Eurasia, the domain of a few loggers, trappers, miners and hunters. Much of the mountain domain is cool beyond the limits of cultivation and is too steep to hold a soil. Upland coolness, which keeps the tse-tse fly and other disease vectors at bay, is, of course, an attraction between the tropics.

Separated by these bare patches, not much sought after for their own sake, the positive components of our distribution are four great clusters. Sixty percent of us are found in the three Eurasian clumps. The largest lies in east Asia, including agricultural China, Korea and Japan. Here people are concentrated on the flood plains and deltas of the mainland and along the southern coast of Honshu. Japan's industrial economy supports a population over half the size of that of the United States on a group of mountainous islands less than the size of Montana. Its example is being followed in Korea, Taiwan and even China.

The south Asian agglomeration occupies the Indian subcontinent. Here monsoon rainfall sustains intense cereal cultivation and a straggle of large cities on the Indo-Gangetic Plain.

The third cluster extends from the Urals to Britain, covering the European peninsula of peninsulas. The region, with a reliable climate from the agricultural viewpoint, has the highest portion of its land under the plough of any continent. Its high yields per acre are, however, produced by less than a tenth of its population. Most people work in manufacturing, services or the information sector, and most dwell in towns.

The newest and smallest population cluster reaches along the Atlantic seaboard of north America from Boston to Washington, D.C., and extends inland through the Appalachians along the shores of the Great Lakes and the Ohio Valley to Chicago and St. Louis. The region has enormously productive soils and considerable mineral resources. Its dense settlement results from European immigration, which quickened in the last century. Farming is efficient and productive; it employs less than one-tenth of the labor force. Most people are town dwellers.

Beyond these four, there are twenty or so lesser foci of population. In the Americas the growth in Florida and the Gulf Coast echoes the earlier surge of Californian population along an axis extending from San Diego to the Bay area. This has its northern counterpart in the Willamette Valley and along the shores of Puget Sound. The southern apex of the Mexican Altiplanicie focused on Mexico City and the basins and valleys of central America have high densities, as do the islands of the Caribbean. There is an axis of high density that runs through the valleys and basins of the northern Andes from Caracas to Quito. Central Chile has a modest density which is focused on Santiago. To the east of the Andes the Rio de la Plata; the São Paulo, Rio de Janeiro, Belo Horizonte triangle; and the coast behind Salvador and Recife house the main population concentrations.

In Africa the lower Nile's floodplain is a strip of great density, as is the Mediterranean and Atlantic coast of the Maghreb. South of the Sahara's emptiness, Kano focuses a sudanic cluster, and a strip of intense farming and cities extends along the west African coast. In the east African plateau the population congregates around Lake Victoria, and in southern Africa the largest numbers are found along the Indian Ocean coast, in the Transvaal and in Cape Province.

In Asia, the Levant, the shores of the Black Sea, the flanks of the Caucasus and the Iranian shores of the Caspian have high densities. Beyond the scattered oases of central Asia, the valleys and coastal plains of southeast Asia point toward the dense sweep of the Malay U, extending from Malayasia through Java to the Philippines. Finally, the empty expanse of Australia is fringed by a thin strand that reaches from Adelaide via Melbourne and Sidney to Brisbane.

Where we are is where the seat of conflict and war lies. The disputed ground and the scene of fighting may be in the relatively empty spaces between us, but the final outcome of major wars is decided in the population core areas of the world.

REFERENCE

Goode's World Atlas. 1990. Chicago: Rand McNally.

4

THE WORLD AT WAR

THE GEOGRAPHY OF WAR SINCE 1945

The apparent demise of Soviet imperialism in the winter of 1989–90, the vanishing popular and political anticipation of a great power conflict breaking out in Europe and the consequent stand down of American and Soviet forces there have all pulled the actual geography of warfare since 1945 into clearer focus. This was immediately reinforced by an outbreak of new conflicts in the world's hot spots.

If we follow Ruth Sivard's example in *World Military and Social Expenditures* and define war as a violent political conflict involving the death of 1,000 people or more, then the only European struggles worthy of note since 1945 have been the Greek civil war of 1945–49 and the Hungarian uprising of 1956. Most of the 127 wars that have occurred since 1945 have been fought outside the urban-industrial world. Asia, with just over half the earth's population, has been the scene of 70 percent of the nearly 22 million deaths that have been caused by wars over the last forty-five years.

During the 1940s (see Figure 4.1), China was undergoing a change of dynasties, and European colonialism had begun its retreat from India and Indonesia. The French hung on in Indochina, Madagascar and Algeria, and Israel appeared on the Middle Eastern scene. In Latin America governments were under attack from the right and the left.

In the 1950s (see Figure 4.2), the Korean War cost three million lives and involved China and the United States and its allies directly. The internal strife in China, Taiwan, the Philippines and Indonesia ran parallel to the anticolonial wars of Indochina and Malaya. The Middle East saw civil war in Iraq and Lebanon and the invasion of Egypt by the British, French and Israelis over the Suez Canal. Anticolonial wars occurred in Tunisia, Algeria and Morocco. Beyond the Sahara, the wind of change began to blow in

Figure 4.1
War in the Late 1940s

Figure 4.2
War in the 1950s

Cameroon, Rwanda, and Kenya. There was civil war in Argentina, Bolivia and Guatemala, and a revolution was successful in Cuba.

The 1960s (see Figure 4.3) saw U.S. intervention in Vietnam, a war which eventually cost two million lives. In Indonesia the government put down an attempted coup and half a million died. The loss of half a million in the Cultural Revolution in China dwarfed the three thousand who died in border clashes with the USSR and India. In the Middle East there was a Kurdish rising in Iraq, strife in the Yemen and the Six Day War between Israel and Egypt. The French finally got out of Algeria. South of the Sahara the Portuguese clung on in Guinea-Bissau and Angola. In Nigeria two million died over Biafra. Tribal strife in Uganda, Rwanda and Zambia was accompanied by violence in Zaire, which was engineered by the copper companies to maintain control in Katanga.

The United States pulled out of Vietnam in 1975, but warfare had extended into Cambodia (see Figure 4.4). The Chinese poked their noses in and attacked Vietnam on behalf of their clients the Khmer Rouge. Offshore there was a Muslim uprising in Mindanao and peasant unrest throughout the Philippines. The Indonesian annexation of East Timor was met with violent objections. The separation of Bangladesh from Pakistan cost the lives of one and a half million people. The Pakistanis succeeded, however, in suppressing Baluchi separatism. The clash between fundamentalists and progressives in Afghanistan had the Soviets intervening on behalf of their clients by the end of the decade. In Sri Lanka there was a leftist uprising. In southwest Asia in the 1970s, the Yom Kippur War pitted Egyptians and Syrians against Israelis. The Syrian government also fomented unrest in Jordan and intervened in the Lebanese civil war. The big turnover came with the ousting of the shah from power in Iran at the end of the decade. In 1974 the Turks invaded the northern part of Cyprus on behalf of Turkish ethnics. Continued political unrest in Turkey itself ended with a military coup in 1980. The Polisarios began their campaign against the Moroccan army in the Western Sahara in 1975. South of the Sahara the last colonial wars ended in Angola and Rhodesia, but newly independent governments were beset with internal stress. Tribal antagonism broke into killing in Burundi, while Uganda suffered under Idi Amin. The struggle between non-Muslim southerners and the government continued in the Sudan. The government in Addis Ababa did battle with Somali irredentism and Eritrean separatism. In Latin America, military rule in Argentina and Chile led to many deaths, and the death toll among suspected dissidents began to mount in El Salvador. In Nicaragua the Sandinistas succeeded in ousting the Somoza regime.

The 1980s brought no respite from war to Asia (see Figure 4.5). The suppression of political unrest continued in South Korea and China. The Chinese clashed with the Vietnamese on their border, and the Vietnamese kept up support of their clients in Cambodia until 1988. Indonesian suppres-

Figure 4.3
War in the 1960s

Figure 4.4
War in the 1970s

Figure 4.5
War in the 1980s

sion on East Timor continued until 1989. The military government of Burma came under leftist attack on several occasions. The Tamil guerrillas commenced their campaign in Sri Lanka in 1984, bringing about Indian intervention. Intercommunal violence continued to plague India. In Afghanistan the Soviets clung on until 1989, but the civil war continued after their withdrawal. For most of the decade Iranians and Iraqis were busy killing a million of each other. The Iranians and Iraqis were also busy killing Kurds. The complexity of violence in Lebanon was compounded by Israeli and Syrian intervention. In Syria the government slaughtered religious dissidents. There was a civil war in South Yemen. The Polisarios pursued their campaign against the Moroccans until 1987. The Ethiopians reached a stalemate in the Ogaden in 1983, but they retreated before the Eritreans in the north. There was rebellion against Islamic rule in the Sudan and a rebellion that brought in the French and the Libyans in Chad. Islamic fundamentalists rose against the government twice during the 1980s in northern Nigeria. In Ghana tribal rivalry broke out into fighting. The Tutsi military massacred Hutu civilians in Burundi. The rivalry of Shona and Ndebele broke out into killing in 1983 in Zimbabwe. In South Africa the police of the white tribe killed 4,000 blacks between 1985 and 1987, and Zulu and Xhosa violence mounted. In the Western Hemisphere, the oddity of the battle for the Falklands in 1982 resulted in only 1,000 deaths, about the same toll as rightist terrorism took in Brazil. In Chile the army killed 3,000 striking miners. The Sendero Luminosa rebels of Peru killed 5,000 soldiers at a cost of 10,000 civilian deaths. The battles of the drug barons in Colombia cost 20,000 lives. In El Salvador civilian deaths of 47,000 compare with army losses of 18,000. The Guatemalan government continued with its savage suppression of rural unrest. The Contra campaign against the government of Nicaragua, encouraged and aided by the U.S. government, cost 25,000 lives. Party rivalry in the 1980 election in Jamaica reached warlike proportions with 1,000 deaths. The number of deaths caused clearly does not discriminate well between political violence, terrorism, civil war, and international war.

POWER AND DISTANCE

Given the age, fertility and affluence of the postindustrial societies of the north Atlantic surrounds, it is unlikely that war will break out there, although some rioting, suppression and terrorism may persist. Fighting with civilian and combatant death tolls in the thousands will continue to occur in the regions of young, growing populations involved in keen competition for power, land and resources. Most wars are being fought at some distance from the great power core areas.

Aside from the significance of economic and demographic trends, there is an explanation for the geography of war which depends upon a postulated relationship between power and distance. In *Conflict and Defense*, Kenneth

Boulding suggested representing the falloff of power with distance from a home base as a loss of strength gradient. The spheres of influence of great powers or, indeed, of any political entities can be conceived of as force fields extending out from some central focus and attenuating in strength with distance from this core area. Power is greatest at home and diminishes with distance from this source. The further the force is extended, the weaker it becomes.

The reason for this gradient is, first, the frictional effect of the cost of overcoming distance. Might weakens with distance because the cost of friction in transporting force and sending command and control messages erodes the magnitude of the energy available. The development in the nineteenth century of naval reach and in this century of radio, air power, rocketry and satellites has diminished this gradient to the point where some would dismiss distance as a factor in the balance of power. The withdrawal of the geopolitical reach of the Soviet empire, however, suggests that distance friction is still a potent force.

Besides the frictional effect of the cost of overcoming distance, the energy required to control a widening expanse of territory comes into play. The cost of hegemonic control clearly increases exponentially with the radius of a power's sphere of influence. The more farflung the bounds of an empire, the thinner spread will be the resources available for its control.

In addition to these material factors, there does seem to be a gradient of boldness for individuals or groups in positions of power. The increasing unfamiliarity which goes with distance from the homeland may well have its effects on national morale and willingness to extend influence. There is a loss of courage or of moral certainty at a distance from home. The boundary between the domains of two powers will be set by a series of thrusts and ripostes, of a diplomatic or military nature, and the resolve of both will wane with distance from their heartlands. Attacks become shallower and shallower until the balance of their courage results in a standoff along some intermediate boundary.

This boundary takes the form of a crush zone or a shatter belt of states and dependencies between power cores. If the level of influence a core area can exert beyond its own boundaries diminishes with distance from the source, then its sphere of influence will be defined by the domain over which its exertable power exceeds that of any other source. If there is some doubt about the power that can be exerted at any distance, as there is bound to be in the case of such a poorly defined force, then, instead of a clear-cut boundary of hegemonic spheres of influence where power levels equate, there will be an area of overlap where it is uncertain which core's influence is dominant. In this interval it is not clear whose power would prevail should it come to violent conflict. This swath of overlapping influences will be the target for persistent political, diplomatic, cultural, economic and indirect military competition. Uncertainty over the outcome may cause hesitation

Figure 4.6
Power Gradients

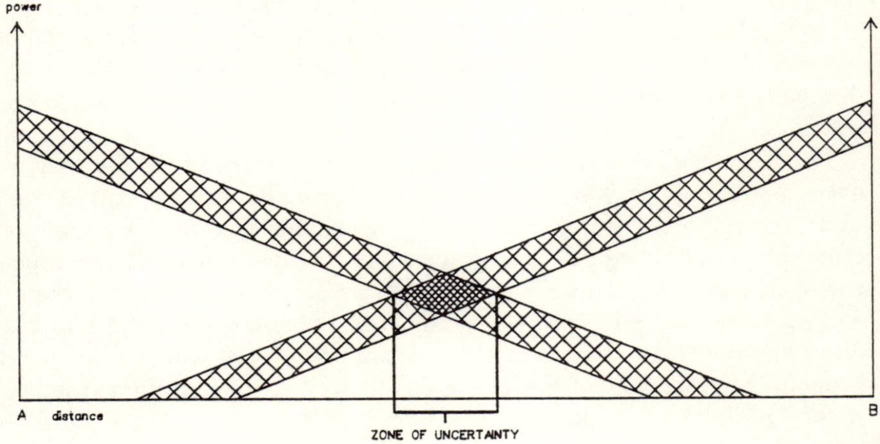

over a direct head-to-head contest for territorial domination here. However, the same uncertainty that generates caution in some may fire ambition in others. War is most likely to break out in the buffer zone or shatter belt because this is where aggressors think they might succeed and where defenders are not certain that they will be defeated.

Although we could imagine the boundaries of spheres of influence being set by a rational appraisal of the lines of intersection of diminishing power force fields without the need for violent competition, too often the reality is war. War, the violent conflict for power over territory, is spawned by ignorance and doubt. An aggressor sees a chance of success where there is some doubt about the equality of power. On the other hand, a defender does not know for certain that he will be beaten. This leads him to make a fight of it, rather than relinquish some of his field of influence. We can represent this condition in Figure 4.6, which shows the power gradients of two competing nations as bands rather than lines.

At any distance from its core, a nation's power potential is known to lie between certain limits, but its effective magnitude could lie anywhere in that range. This produces an overlap of spheres of influence and a zone of uncertainty. In this interval of the space between the poles of power there is doubt about who could exert the greatest force. This ignorance, especially at the outer fringes of the fields of dominance of great power, creates the temptation and opportunity for local enterprise in war, where the hegemonies' power is weakest and the uncertainty as to who is the most potent is greatest. Radical adjustments in power gradients, like the present Soviet pullback, add to the confusion, for no political or diplomatic precedents set the new upper and lower bounds of exertable power for the global or regional hegemonist. Such new limits are currently being groped for at the head of the Gulf.

Even outside the zone of overlapping great power potential, uncertainty provides the occasion for war. Dissidents may seek to overthrow the incumbent government in a country lying clearly within the sphere of influence of one great power when it is not certain that their ability to achieve their ends with force is less than that of the government, or the dominating power, to withstand them. Despite the location of trouble well beyond their direct military reach, the distant hegemonist will be tempted to interfere indirectly. Here is an opportunity to discomfort the competition in its own backyard. This carries little risk as long as the prospect of direct confrontation is avoided. Thus, such insurgents often get encouragement and clandestine aid from a great power on the other side of the earth, as did the Mujahedin of Afghanistan in the 1980s.

This, then, is the geostrategic setting for most present-day wars.

CURRENT HOT SPOTS

The events of 1990 have proven beyond a shadow of a doubt that this is an uncertain world where prediction is a foolish enterprise. This would suggest some hesitation in offering even a description of the present state of global conflict. It will be dated before the ink is dry. Waiting for January 15, 1991, was enough to humble any would-be prophet, and it would not be surprising if my geographic snapshot as of the end of January 1991 proves to miss the point.

Africa

Southern Africa. In 1838 when the Dutch farmers defeated the Zulu at the battle of Blood River they laid the foundation for their domination of southern Africa. After their defeat by the British in the Boer War they built up their political apparatus, and from 1948 their Nationalist Party ruled and imposed their racial dominion over land and resources in the form of apartheid. The territory of the white-ruled republic was interrupted by the geographical expression of separation in the tribal homelands and kingdoms, while the Group Area laws created segregated enclaves in the shantytowns and townships on the fringes of urban-industrial South Africa. Sixty-eight percent of the population was confined to 13 percent of the agricultural land or held at arms length when they were needed in the towns and mines. The police and military force necessary to maintain this system internally and to defend it against insurgents living beyond its borders, or an antiracist holy war mounted by the frontline states to its north, made a considerable call on the resources of the white population who constitute less than one-fifth of the total. The predominantly white army that can be mobilized is twice as large as the combined armies of the seven black nations to the north and is much better equipped. Nevertheless, in the late 1980s the cost in money and lives of military operations in Namibia and Angola was beginning to tell.

The other time-honored device used to maintain Dutch dominance was the divide and rule strategy. Now as the draconian controls that sustained white supremacy are dismantled, the tensions it employed are breaking out into communal warfare within and between ethnic groups, with a mounting death toll. In 1988 the puppet governments of the tribal homelands of Transkei, Ciskei and Bophuthatswana were toppled by supporters of the African National Congress (ANC). This party, which was banned until 1989, found its main strength among urbanized Xhosas. Its leader, Nelson Mandela, released from confinement by President de Klerk in the hope of achieving a peaceful, political path of change, is a member of the Xhosa royal clan. The recruiting of Zulus in the townships of Natal into the ANC had been met with resistance by members of the Inkatha movement, which turned to killing in 1983. Inkatha was an established ethnic movement led by Mangosuthu Buthelezi, the chief minister of the Kwazulu homeland. The Zulu account for the largest portion of the population, about a quarter of the total. The fighting and killing in the streets of Durban and Pietermaritzburg continued through the 1980s. Inkatha had spread its reach beyond Natal into the Transvaal industrial areas, where Zulus had migrated to work. With the legalization of the ANC, the intensity of fighting increased in Durban and spread to the Transvaal, until over 5,000 were dead. The government has made some attempt to stop the killing, but there is more than a suspicion that many of the police, who are opposed to the demolition of apartheid, have encouraged violence to discredit the ANC. Although the fighting is mostly between Zulus, it does feed on a strong historical antipathy between Zulu and Xhosa. Many ANC leaders are overtly anti-Zulu. As the politicians seek a constitutional compromise and white extremists propose a white homeland solution, violence rages among the apolitical young of the township and shantytowns with little to lose but their impoverished lives. Forty-two percent of young blacks are unemployed.

Ethnic conflict was sponsored by the South African government beyond its borders to keep potential foes off balance. With the unravelling of apartheid there has been an attenuation of this interference and a relaxation of tension in some of the frontline states. In Mozambique the struggle costing 19,000 lives since 1982 between the government in Maputo and the Resistencia Nacional Moçambicana (RENAMO) rebels, subsidized by South Africa to disrupt the railways in particular, appears to be coming to a close with negotiations in Rome. In Zimbabwe the South Africans supported Ndebele guerrillas in their fight against Shona domination under Robert Mugabe. Now Mugabe and the Ndebele leader, Joshua Nkomo, have made a pact and the state of emergency has been lifted. Even though the celebration of Namibian independence in 1990 was marred by a gunfight between Southwest Africa People's Organization (SWAPO) rebels and South African troops, its first year has remained calm. The political settlement in Namibia has been followed by the pursuit of an end to fifteen years of war in Angola. The government controlled the area around Luanda and every March Jonas Savimbi's Uniao Nacional para a Independencia Total de Angola (UNITA) forces pressed from the south, usually severing the Benguela railway and the Atlantic exit for Zambia's copper production. The 1990 campaign was followed by a ceasefire in December as a prelude to talks prompted by the United States and the Soviet Union, to be held in Lisbon between Savimbi and the government.

Equatorial Africa. In equatorial Africa President Mobutu, impressed by the execution of the Ceauşescu in Romania, raised hopes of a multiparty system in Zaire and triggered a surge of hope and unrest. In May 1990 his army savagely put down those suspected of harboring separatist ambitions in the copper-rich southern province of Shaba (Katanga). There and elsewhere resentment smoulders. Mobutu has maintained the upper hand in Zaire since 1965 by de jure abolishing the tribal structure, but de facto allowing the chiefs to rule their homelands. The central government and modern sector are largely urban and the government can exert power with planes and paratroopers if necessary.

In the highlands that overlook the Congo basin from the east, strife whose seed was sown 400 years ago continues to plague Rwanda and Burundi. The tall, pastoral Tutsi, who conquered the Hutu farmers maintained their lordship by slaughtering any dissidents. In Rwanda the Hutu overthrew them and gained the upper hand in the 1960s. In Burundi, however, the Tutsi remain the rulers and soldiers, although they constitute only 15 percent of the population. In 1972 the Tutsi killed 100,000 Hutu. In the summer of 1988 they killed at least 5,000 more; 35,000 others fled into Rwanda. Meanwhile Tutsi rebels, who had sought refuge in Uganda, invaded Rwanda at the end of 1990. In Uganda itself the factionalism of Baganda, Kakwa, Acholi and Lango, which reached a bloody crescendo under Idi Amin's misrule, lies not far beneath the surface waiting to erupt again. In Kenya the Kikuyu had been cautious not to appear to be too dominant, but efforts to establish an opposition party in December 1990 were met with sharp suppression. In Tanzania, where German colonial administrators had suppressed the tribal structure, ethnic unrest began to show itself in the 1980s. The tyrannical rule of descendants of American slaves, which had persisted in Liberia since the 1840s, was overturned in 1980 to be replaced by tribal strife for power with Ghio and Mano fighting Krahm and Mandingo. There are three rival presidents.

The Islamic Borders. In a swath across Africa from Gambia to the Horn and south along the coast to Dar es Salaam lies a shatter belt between the Arab and Bantu worlds. Here Islam and black Africa meet and interact.

The Tanzanian island of Zanzibar was an Omani slaving colony in the last century. In 1964 when Arab rule was finally overthrown, 5,000 Arabs and Muslim Indians were slain. In Somalia President Barre maintained the upper hand since 1969 by playing Muslim clans off against each other to get dissidents removed. In December 1990 he made the mistake of alienating the Hawiye, the large clan that dominates the area around Mogadishu. They seized the city and chased Barre and his small Marehan clan from power. To the northwest in their highland fastness, the Amhara are being driven back from the outer reaches of their empire by Muslim and non-Amharic peoples. In the north the Ethiopian army lost control of Eritrea in the mid-1980s. In March 1989 they were pushed out of Tigre province. Separatist rebel forces advanced within reach of Addis Ababa as the Soviets untangled themselves from their last African client, Mengistu Haile Mariam. From beneath his Marxist veneer Amharic imperialism begins to show more clearly. Aid has been dispatched from Addis Ababa to the non-Muslim southern Sudanese who have been fighting against the imposition of Islamic sharia law by the soldiers who grabbed power in Khartoum in mid-1989. After bombing southerners in December 1990, the government has refused to acknowledge famine in the pagan and Christian south or to facilitate relief. The conflict in Chad between black Christian southerners and the combination of Tauregs and Muslim blacks in the north continues, inviting the interference of the French and the Libyans. In December 1990 a Muslim warlord with Libyan support chased the prior ruling warlord off to Cameroon, while the French stood aside.

In Nigeria the black Muslims of the north arrived as conquerors from the Sahara in the nineteenth century. The British, with characteristic colonial inconsistency, muddled together in Nigeria a combination of indirect rule using the Muslim structure in the north and the direct rule of the pagan and Christian south. This difference planted the seed of Biafran ambitions to secede in the 1960s, which resulted in two million deaths. The tension over sharia laws survives as a major item of politics. In the 1980s 5,000 people died in Muslim versus Christian rioting. In 1988 in Kaduna, for example, there were riots and killings when the Muslim majority reacted to Christian profanity. In Senegal and Mauritania fighting broke out in 1989, leaving light-skinned Mauritanians dead in Senegal and black Senegalese peasants dead across the border. The Mauritanian army has been busy killing and torturing blacks. Before the French colonial era, Arabic- or Berber-speaking Moors dominated and enslaved blacks in this part of the world. In Mauritania there is a deep divide between the Moorish herders and the black Muslim farmers, the Fulani and Wolog. Here racial conflict divides the Muslim camp. Slavery was officially abolished in 1980, but it is still practiced. The Beydene, the white soldiers who rule the country, and their Arabizing influence is resented and resisted. Blacks rule in Mali, but Taureg nomads harass the government in the northern regions.

North Africa. The war in the Western Sahara was among Arabs. The Polisarios are nomadic Saharawi, or bedouin Arabs. They contested with the Moroccan army for sovereignty. The Polisarios are now left high and dry because their patron, the Algerian government, has lost its revolutionary ardor and has embraced Morocco, monarchy and all. Far beyond this exercise in Moroccan imperialism as a source of disruption in world affairs has been Colonel Qaddafi's campaign to promote Libya's influence. This has pushed south into Chad and has aroused west African fear of his geopolitical ambition to create an Islamic corridor across the Sahara to the shores of the Atlantic. That fear seems to be misplaced. Even if the ambition exists, the task would be too much for Libya's forces. Qaddafi's bid for leadership of the Arab world has been eclipsed for now by the activities of Saddam Hussein, and he has to content himself with the role of peace broker. The Maghreb Union, formed in 1989 to increase the economic interaction of Morocco, Algeria, Tunisia and Libya with a view to eventual political union, may have relaxed some stress. The tension between Berbers and Arabs certainly seemed to have abated until the government made compulsory the use of Arabic in Algeria. The strain shows in the rapidly growing cities of the region. Heavy unemployment and poverty among a young population generate heat that might reach flash point. Among the increasingly urban poor of North Africa, Islamic fundamentalism builds a following and threatens the status quo. Riots rocked Morocco at the end of 1990 as shantytown youths attacked the city centers of Fez, Tangiers, Rabat and Agadir and were met with a military response. In Egypt the Islamic brotherhoods have their own soldiers, the *jammat islamyia*, who launched a campaign of attacks on Americans and officials in Cairo in 1987. Unless they can penetrate the upper ranks of the army, however, they have little chance of success.

Asia

The Middle East. We still attach the British imperial designation "Middle East" to southwest Asia. This was the Middle East Command, reflecting its global position with respect to London. If its significance was in standing astride the passage to India while Britannia ruled the waves, it has been restored fully to its geopolitical centrality since the British withdrew from east of Suez. The pivot of the continents, where Africa and Eurasia meet, where civilization had its beginnings, where empires clashed, it has been the focus of conflict through the ages. In this century the disputes between farmers and herders, between cities, between empires, and between Arabs, Turks and Iranians have been made immensely more complex by the discovery and exploitation of oil and the presence since 1948 of Israel. Clearly the urban-industrial world's governors will not let local fights take their course, nor leave local lords unattended. The oil companies call the shots now, even if they did not rack the balls.

Although the present rulers of Egypt have called it quits with Israel, Israel's existence is deemed a geographical affront by most Arabs, a Western colonial outpost in their midst. Those Arabs within its borders feel this most strongly, especially now with the prospect of a million Jews arriving from the collapsing Russian empire, increasing the population by a fifth. Since Arabs still make up 15 percent of the population, Israel's internal security involves an enormous outlay of people and structural defenses against dissidents, which adds to the burden of external defenses. The internal structure came under heavy attack with the *intifada*, the rising among Palestinians in Gaza and the West Bank which began in December 1987. Since that time 790 Palestinians have been killed and 57 Jews. The security forces' heavy handedness and violent reaction, especially on Jerusalem's Temple Mount in October 1990, diminished sympathy for Israel for a while. Scud attacks from Iraq, aimed at bringing Israel into the Gulf War and breaking away Arab states from the coalition, however, rekindled the West's affection for Israel and led the West to provide Patriots for its defense against the missiles. Israel's Arab population smolders under curfew and stringent guard. The external potential for an attack against Israel through Lebanon, Syria, Jordan and Egypt is faced with a nation in arms with short interior lines and a powerful air force. Short interior lines also, however, imply little defensive depth. Any Israeli offense, either before or after an attack through Lebanon, across the valley of the Jordan, or through the Gaza Strip, must take into account the difficulties of operating mechanized forces in a densely built-up urban landscape.

The densely occupied lowland corridors of Lebanon have been torn apart by warring factions, and the nation is divided into occupation zones, each of which, in turn, is fragmented into fiefdoms. The Israelis have kept a security buffer in the south since 1985, leaving it in the hands of their Christian, South Lebanon Army allies and use it to shell Shia villagers in the Bekaa suspected of harboring fundamentalist guerrillas. The Shias control most of the south where Amal and Hezbollah militias battle for dominance, as they do in Beirut. The Druze control the area to the south of the city, and the Maronites of the Phalange control its northern approaches, having pulled out of the city itself at the end of 1990. The north and east of the country, beyond the mountains of Lebanon, are commanded by the Syrian army, including the maronite canton ruled by Suleiman Franjieh. The whole country operates under a regime of intermittent violence, like a set of fiefs ruled by robber barons who are occasionally constrained by their feudal overlords.

Apart from its irredentist ambitions in Lebanon, hatred for Israel, and desire to regain the Golan Heights, the government of Syria is concerned about Turkish dam building on the Euphrates and leadership of the Arab world. Control of the river may be a matter for negotiation, but the competition for regional hegemony drew Syria into the Gulf War. Both Hafez Assad of Syria and Saddam Hussein of Iraq are Baathist, members of a party based on the secular ideals of Arab nationalism and socialism. They are rivals for Arab leadership. They must both contend with the battle between Sunni and Shia which has riven Islam since its foundation. Baathism came to power in the hands of minorities in both countries—Alawites in Syria and Sunni in Iraq. Islamic fundamentalism has found a following among the excluded majority, and both rulers have been ruthless in suppressing such dissidence. Hafez Assad let his brother loose with the army on the Sunni city of Hama in 1982. The underground Dawa movement that made headway among the Shia in Iraq was scotched in 1980 with a number of executions. During the war with Iran, the Shia population of the south decided that they were Iraqis first, just as the population of Khuzistan, the oilfield area of Iran at the head of the Gulf, had decided that they were Iranian nationals before they were Arabs.

Saddam Hussein replaced his opponent of eight years of war over the Shatt al Arab with a far-flung alliance mustered under the command of the United States, which includes Hafez Assad of Syria. Hussein achieved this by overstepping the military and moral mark in his conquest of Kuwait. The Iraqi claim on some Kuwaiti territory, which sought an adjustment of the borders created by the British for a puppet emirate in 1962, was reasonable. The Iraqis had cause to be aggrieved by Kuwaiti and Saudi breaches of oil output agreements, if not by the excessive Kuwaiti pumping from an oilfield cut by the border. There is a good case to be made that Saddam Hussein was encouraged in his aggression by what he might have taken as a lack of opposition from the United States. Just as Stalin and the North Koreans calculated on the basis of Dean Acheson's exclusion of the Korean peninsula from the American defense perimeter in 1950, so Saddam Hussein was misled by the U.S. ambassador to Baghdad's telling him that the United States did not wish to become involved in Iraq's border dispute with Kuwait. On July 31, 1990, the deputy secretary of state for the Middle East announced in public that the United States was not committed to defending Kuwait if it were attacked. Only after the event were actions taken to oppose a man who is known to carry out his threats. Iraq succumbed to the air superiority of the coalition and retreated from Kuwait in a rout. Saddam Hussein then turned what was left of his forces loose on the Shias in the south and the Kurds in the north.

The Gulf War can be construed as a belated American effort to fill the geostrategic vacuum left by the British when they struck camp there in 1969. The failure of the United States to step into the breach then left it open for the Soviet Union and local aspirants to build up their power bases. With the collapse of the Soviet empire, the United States is left to contend with the locals for dominion over this pivotal turf. The Gulf War is a massively amplified replay of the British flying troops into Kuwait in 1962 to forestall an Iraqi threat of invasion. The Iraqi revolutionaries, who had chased the Hashemite puppet king from the throne in 1958, sought to realign the British-imposed borders.

With half a million American troops along its northern borderlands, it is plain that Saudi Arabia's theocratic, Sunni monarchy is sustained by the support of the United States. Apart from Saddam Hussein, the Saudi status quo is troubled by the Shia population of its eastern oilfield region, by the Hajjis who flock to Mecca annually, and by those who stay to swell the growing proletariat of Jiddah. In the southwest of the Arabian peninsula, the joining of the two Yemens into a republic of 13 million people in May 1990 has caused the Saudi autocracy some unease.

The Northern Highlands. The Zagros and Toros mountains separate the Arab world to the west and south from the plateau dwellers of Turkey, Iran and Afghanistan. The northerners are predominantly Muslim but ethnically distinct. The Turks are drawn between Europe and Islam. They have a geographic foothold in Europe in Istanbul and Thrace, and they are founder-members of the North Atlantic Treaty Organization (NATO). They use Latin script and a European code of law. They are negotiating membership of the European Economic Community (EEC). Turks have provided industrial Europe with much of its *Gastarbeiter* supply of labor. Turkish nationalism, however, still runs deep. The Turkish army invaded northern Cyprus in 1974 to protect the Turkish population there. Though the government was cautious over Caucasian fighting between Armenians and Azeris in 1989, it is clear where the sympathies of most Turks lie, and this extends to the Tartars, Uzbecks, Kazakhs and other Turks of the Soviet empire. In the summer of 1989 Turkish nationalism was further aroused by the expulsion of thousands of Turks from Bulgaria.

The Turks share a minority in common with Syria, Iraq, Iran and the Soviet Union. The homeland of the Kurds is divided among these states: 9 million in Turkey, 0.6 million in Syria, 4 million in Iraq, 4.5 million in Iran and 0.2 million in the USSR. These martial mountain dwellers, with their Indo-European language and Sunni faith, provided Islam with its greatest warrior hero, Saladin. Stirring one's rival's Kurds is a time-honored game in this part of the world. The Kurds have usually had contentious relations with governments. Most recently they got caught up in the war between Iran and Iraq, and in Iraq they paid the price of Iranian support, extracted with poison gas and destructive resettlement schemes by the forces of Saddam Hussein. After Desert Storm, they are once more being harried. In Turkey, after Kurdish guerrillas killed fourteen Turks in 1987, the army attacked and employed the right of hot pursuit granted them by Iraq in 1984 to hit Kurdish bases in Iraq from the air. The massing of Turkish forces on the Iraqi border was interpreted by some to be not merely anti-

Kurdish but to be covetous of Iraq's northern oilfield. Turkish cooperation with the anti-Iraq coalition adds to this suspicion.

The Iranians are kin to the Kurds linguistically, but they ruthlessly manipulate the Kurdish warlike nature for their own ends. In 1974 the shah encouraged a Kurdish rebellion in Iraq but, the following year, he reneged on his promise to the rebels for a share of sovereignty over the Shatt al Arab. The Kurds were enthusiastic Islamic revolutionaries in Iran in 1979, but their desire for a degree of autonomy was violently suppressed by Ayatollah Khomeini's Revolutionary Guard in 1984. The Iranians then encouraged rebellion among Iraqi Kurds during the war with Iraq but left them to the tender mercy of Saddam Hussein after the July 1988 ceasefire.

Near defeat by Iraq and the death of Khomeini left Iran quiescent. In the past the Iranians have squabbled with the Turks over the Kurds', now the two countries share a common interest in a growing trade arrangement. The Iranian leadership continues to preach Shia jihad against the Sunni and the West. Iranian Revolutionary Guards and support are provided to the Hezbollah faction in Lebanon. Although they condemned the Soviet presence in Afghanistan, the Iranians subsidized only Shia Mujahedin, who killed Sunni Mujahedin as readily as Soviets. Having been bribed into neutrality by a final settlement of the Shatt al Arab dispute by Saddam Hussein in November 1990, courted by the Soviets and the Americans, the mullahs continue to rule in Teheran with little opposition. In time the bourgeoisie may grow and press for freedom from clerical control and an expansionist economic policy. If this infects sufficient numbers and the officer class, then an internal contest for power might ensue, but 5,000 executions in three years may deter activism for now.

In Afghanistan the mullahs have not got the upper hand yet, despite the withdrawal of the Soviet army. The government in Kabul has withstood the attacks of the Mujahedin. A push, before the snows of winter and before Pakistani and U.S. aid dried up, in October 1990 by Gulbuddin Hechmatyar's fundamentalist group seized the small town of Tarinkot when the garrison of 500 joined the rebels. Kabul, replete with mine fields, missiles, tanks and artillery is a different proposition, and an attack on the capital failed. U.S. and Soviet diplomats are evidently looking for an acceptable power sharing arrangement between Mohammed Najibullah's government and the rebels. Selling a deal to the Mujahedin, fragmented by internecine strife, will be another matter; in January 1991 they attacked Kabul with rockets, killing eleven civilians. When the snow melts the tempo will pick up.

The Indian Subcontinent. To the east of the Hindu Kush in Pakistan the electoral victory of the Islamic Democratic Alliance and the defeat of Benazir Bhutto in October 1990 brought politicians representing conservatism and the urban middle class of the Punjab to power, with the approval of the army. Pakistan is still at odds with India over Kashmir and tension mounted in the spring of 1990, but troop numbers along the border were reduced in the winter. Having been fought over twice since the end of the British raj, the current boundary is the cease-fire line of 1965. Pakistani Kashmir is overwhelmingly Muslim; Jammu and Kashmir in India is 60 percent Muslim. Muslim secessionists took to guerrilla fighting in January 1990 after fifty demonstrators were shot by Indian police, and they have refused to negotiate. The Indian prime minister instituted direct rule from Delhi and threatened war if Pakistan did not desist from inciting an uprising. The Pakistanis remained calm but prepared, and diplomats continued to talk while Muslim refugees crossed the border into a Pakistan already overburdened with Afghanis. In the far south Pakistani unity faced the test of trouble in the Sind. In February 1990 shooting broke out between police and dissidents in the streets of Karachi. These rebels, from the Mohajir majority in the city, are descended from Muslim refugees who arrived from India in 1947. The indigenous Sindhis resent that this Urdu-speaking population dominates their chief city. The defeat of Bhutto, whose power base lay, with her estate, in the Sind, increases the prospect of a violent Sindhi separatist movement, aroused by the prospect of Punjabi dominance.

Passing from the valley of the Indus to the Ganges, India presents a plethora of resentments and smoldering embers, apart from Kashmiri Muslims. In the Punjab, Sikh fears of being swamped by Hindus broke out into killing, averaging eighteen deaths a day in 1990. Downstream in the heartland of Uttar Pradesh, violence broke out in November 1990 in Ayodhya where militant Hindus wished to tear down a mosque to build a temple devoted to Rama in his birthplace. This sparked clashes between Hindus and Muslims throughout the country and caused a thousand deaths. In Maharasthra, the hinterland of Bombay, there has been intermittent violence between Hindus and Muslims for years. Assam presents a welter of antipathies among tribal folk, the longstanding Hindu Assamese population, descendants of preindependence Bengali Muslim immigrants, and now illegal Bangladeshi Muslim refugees from poverty. Students from the tribal groups first turned to violence in 1985, and the influx from Bangladesh keeps up the pressure. In the eastern hill country, Sino-Tibetan, tribal Nagas want to separate themselves from India. The government imposed direct rule from Delhi in December 1990. Along the zigzagging McMahon line separating India from Tibet in the Himalayas, Indian and Chinese soldiers have sniped at each other since 40,000 Indian soldiers died in a war here in 1962. Soviet Premier Gorbachev engineered the beginning of a rapprochement in 1988 when he persuaded Rajiv Gandhi to visit China as the first Indian prime minister to do so since his grandfather went in 1954. A deal involving Indian cooperation over Tibetan dissidents and refugees was bandied about.

Strife also exists in the Himalayan kingdoms. In Bhutan Hindu Nepalese emigrants are overwhelming the local Buddhist Drukpas, and their reaction has flashed into violence. In Nepal, after some police killing of demonstrators, the king has backed down in the face of widespread demand to widen the circle of power, and he has accepted a democratic constitution.

There is now a nearly universal, rather than a regional, source of violence in India in the upper caste unrest over the affirmative action policy undertaken to improve the lot of the 44 percent of the population who come from the "backward" castes.

In Tamil Nadu there was great dissatisfaction over the Indian army's employment against the Tamils in Sri Lanka. Hindu Tamil demands for greater power and autonomy in northern Sri Lanka have been met with resistance by the government who represents Buddhist Sinhalese indigenous interests. In over seven years of fighting, 14,000 people have been killed. After the withdrawal of the Indian army, the Sri Lankan army continued to fight the Tiger guerrillas. In October 1990 they staged a dramatic relief of soldiers besieged in the 400-year-old Jaffna fort. Abandoning the fort as militarily useless, the army failed to get control of the whole Jaffna peninsula and helped Tiger recruiting by a policy of indiscriminate bombing in the war zone. After a brief cease-fire, the slaughter resumed on both sides at the end of January 1991.

Southeast Asia. In 1988 a group of generals seized power in Burma and changed its name to Myanmar. They have refused to hand over power to duly elected politicians. They fear that a peaceful settlement with tribal dissidents in the northern, forested mountains will make many of the 200,000 soldiers in this nation redundant. Their fear of the democratic movement found expression in a program of urban renewal for the capital Yangon (Rangoon) instituted in 1990. Half a million people were ejected from the city to break up neighborhoods with a suspect population. The northern guerrillas do have a source of revenue in their poppy crop. The traffic in poppy juice is in the hands of remnants of the Kuomintang who arrived in 1949.

Thailand's military elite indulged in twenty-six attempted coups in twenty-seven years, having the luxury of a universally accepted, figurehead monarchy which provided consistency and stability to the nation. Apart from competition among military factions for power, the main sources of danger for Thailand lie on its borders. The government chooses to leave the remains of Malayan communist insurgency to their own devices along the border with Malaysia. Since 1985 the Khmer Rouge, who escaped the Vietnamese capture of Ampil in Cambodia, have added to the confusion of the Cambodian refugees along Thailand's southeastern border. Their presence sparked a series of border incidents that lasted through the late 1980s. With the withdrawal of the Vietnamese army from Cambodia and the drive by the Khmer Rouge for Phnom Penh, this pressure increased with 150,000 refugees fleeing war and poverty. In January the Thai government began to close the camps. After eleven years in the jungle, Pol Pot and his Khmer Rouge, in alliance with the forces of Son Sann and Prince Sihanouk, are making a bid for power, pushing along the road from Bangkok to Phnom Penh. The government holds the towns and has instituted a "strategic hamlets" policy to deny the insurgents food. The rebels move through the countryside at will, sowing mines, probing closer and closer to the capital without providing government forces with sitting targets. With only 25,000 people, the Khmer Rouge have little chance of winning a conventional war. The government army is not very good, but it is twice as large. Sihanouk and Hun Sen, the Cambodian prime minister, met for talks in Paris in December 1990. Whether the Khmer Rouge would respect any agreement is another matter.

Having withdrawn from Cambodia, the Vietnamese are at loggerheads with China over the Spratly islands and their oil prospects. The elevation of General Giap to the presidency is hardly likely to reduce their militancy. There was an exchange of gunfire over the islands in 1988. The Malaysians and Filipinos also lay claim to the islands and, if there proves to be much oil, these islands may be the cause of a many-way fight in the South China Sea.

Apart from this, the Malaysians have a potential bone of contention with Indonesia in their eastern oil-producing states of Sabah and Sarawak on the island of Borneo, for which they fought a war in the 1970s. Internally Indonesia faces the unsettling prospect of President Suharto's demise which could generate internal and external strife. When Sukarno fell in 1965 half a million died in the chaos, and there were firefights with Malaysia and Singapore. The Indonesian government has dealt severely with resistance to Javanese colonialism in East Timor, West Sumatra and West Irian. They have also extended a fraternal, Islamic hand to the Muslim separatists in Mindanao, the southernmost island of the Philippines, with a Muslim population of five million. The Christian majority of the island's population are not too keen on granting them autonomy. Throughout the rest of the Philippines insurgency among an impoverished tenant peasantry is widespread. Out of a total of 42,000, 7,000 villages are under rebel control. Rebels operate within two miles of President Aquino's office. The army has undertaken a somewhat dirty campaign of harassment, kidnapping and assassination and has added to the corruption of the forces of law and order, which found expression in the attempted military coup of December 1989. The army has achieved some success, reducing guerrilla numbers from 28,000 to 20,000. But in many provinces government forces are an army of occupation and sympathy lies with the rebels, especially now that they have eschewed their doctrinaire Marxism. In response the government has disbanded the Constabulary, Marcos' death squad, and has reduced the power of the army. The prospect of U.S. withdrawal of its military presence from Subic Bay and Clark Air Force Base would, perhaps, leave the government of Corazon Aquino less secure.

In Melanesia the somewhat brutal takeover by Fijian soldiers, in the face of dominance of the economy and politics by the Indian half of the population, is understandable if inexcusable. British influence has failed to bring about a compromise even yet. Colonel Rabuka has proposed a constitution that would prevent the Indians from regaining power. The French colonial presence in New Caledonia—one-third of the population are expatriate French involved in extracting iron ore, nickel, chrome and manganese—upset many of the indigenous Kanakas. The expedient of partition was established in 1988. The Micronesians have achieved a variety of degrees of independence from the United States. Some would like the U.S. military presence on Guam removed. The Polynesians, having been overwhelmed in Hawaii and New Zealand, are now being swamped by tourists in a seasonal form of colonialism.

The Far East. A power vacuum has been created in the Far East by Soviet collapse and the start of American withdrawal. Japan has extended its commercial empire across much more than the western Pacific and Southeast Asia. There are disturbing signs of growing militancy and hankering for military glory within Japan and for occasions like the Sakhalin and Kuril islands dispute with the USSR and the three-way spat with China and Taiwan over the Senkaku or Diaoyutai islands, just northeast of Taiwan. The vast majority of Japanese still favor a pacific role for their nation in the world, and the Soviets have agreed to come to terms on the Kurils.

Opposite Japan, the rugged Korean peninsula has been divided by a mile-wide demilitarized zone since 1953. There are bristling defenses and repressive, militaristic governments on either side, and reunification is a constitutional mission of both. The growing affluence and nationalistic self-confidence of southern society is irked by the continuing American direction of its defenses, and it looks forward to U.S. withdrawal in 1992. Since the Soviet government has made its peace with China, North Korea's geopolitical significance has vanished, and the USSR has established diplomatic relations with South Korea. This left the Kim dynasty in the north isolated but vulnerable to the attentions of China and Japan. Talks have been held between North Korea and South Korea, but the apparatus of violence which has stood ready for forty-five years is by no means defused.

In China the internal stress set up by Deng's redirection of Chinese energies along the lines of hedonistic individualism, contradicting not only the communist political structure but also the social traditions of China, was on view in Tiananmen Square in 1989. There is a growing regional stress between the interest of the increasingly wealthy in the booming, coastal provinces and those in what is in some cases dire poverty in the interior and west. Among the non-Han peoples, the Tibetans continue to chafe under Chinese rule; Turkic Muslims in the northwest may be caught up in the turbulence of the collapse of the Soviet empire in central Asia. Externally China is no longer so threatening, and the reestablishment of relations with the Soviets, marked by Gorbachev's visit to Beijing just days before the Tiananmen massacre, went almost unnoticed. There has even been an exchange of pleasantries with the Mongolians. The Taiwanese declared an end to the state of war with China which has prevailed since 1948. To the south, Vietnamese economic needs and their withdrawal from Cambodia have closed the gap of animosity with China there.

USSR

In the winter of 1990–91 Premier Gorbachev abandoned the objectives of economic and political reform in favor of preserving the union and law and order. He put together a centralist government with the backing of the military-industrial complex and the might of the KGB and army to resist rebellious republics. The draft was the issue to test the sovereignty of the republics. In the Baltic republics of Estonia, Latvia and Lithuania the centers of information and control were seized with some loss of life. Meanwhile, the four largest republics, Russia, Belorussia, the Ukraine and Kazakhstan, threatened to wreck Gorbachev's union treaty, which maintained some central authority while offering autonomy in all matters except currency and defense. These four republics propose to establish a defense treaty among themselves. The chiefs of the army and KGB responded by conjuring up fear of bloodshed and civil war should the Soviet Union break up. If the army and KGB attempt to impose their will, there well may be war, especially in central Asia where Islamic revivalism and memories of resistance to Bolshevik centralization in the 1930s are strong. Along the flanks of the Caucasus, which the Islamic and Christian worlds share as a fragmented buffer zone, violence broke out in 1989 between Armenians and Azeris. In January 1990 the army killed 125 in Baku while quelling attacks on Armenians. By August Azerbaijan authorities had erected a full-blown border with Armenia, and Armenian guerrillas were carrying out raids across it on behalf of their kin in Nagorno-Karabakh. In April 1989 the army killed twenty people in Tbilisi, the capital of Georgia, to stop prodemocracy demonstrations. A year later the local police looked on while crowds trashed the KGB headquarters in the city. In the Moldavian republic, the Romanian majority are inclined to seek reunion with Romania, while the Russian speakers recruit the Turkic, Christian Gagauz to the cause of Soviet unity. There are 60 million Russians outside the Russian republic. These add to the tension and possible sources of conflict. In all of this turmoil and confusion, there is the likelihood of localized killing at least as the Russian empire crumbles after a turbulent century of existence.

Europe

With Mother Russia pulling back her skirts from Eastern Europe, the Central Front, extending across the continent between East and West Germany, vanishes as the world's premier line of confrontation. The soldiers and gear located here, in what was the possible site of an escalation to all-out, nuclear war, have been rendered redundant. Although the Russian Republic on her own would still constitute a major power, her government's attention and energy are drawn to the needs for internal reform. What has been unleashed is the fissiparous force that acts along ethnic fault lines and is so evident within the former Soviet empire. Since lines of cultural cleavage do not coincide perfectly with state boundaries, there are a slew of potential territorial disputes between nations, as well as sources of regional rifting within them. These regional tears are most marked in the upland areas where physical geography has conspired through history to preserve cultural diversity in mountain refuges, while on the plains sweeping changes have wrought large areas of linguistic homogeneity. Even so, on the plains, political lines have been drawn which cut across the disposition of similar people, creating grounds for dispute.

Romanians have cast an irredentist eye across the border, to the Moldavian republic, and, once they have sorted themselves out at home, they may do so again, despite its significant Russian population. The spark that caught and fired the overthrow of the Ceausescus was struck among the two million Magyars of Transylvania. The Romanian government had taken exception to the cultural propaganda campaign the Hungarians had directed at Transylvania. This became one of the key targets for Ceausescu's "systemization" of the landscape and eradication of villages around Timisaoro. Hungary and Romania nearly went to war over this in 1988. Animosity between Vlachs and Magyars in Transylvania remains and finds expression in antigovernment protests in Timisaoro. South of the Danube the Bulgars have been content with being nasty to their Turkish population and driving them over the border. To the south of the Rodophe the Greeks and the Turks are still at loggerheads over the Aegean islands, Thrace, and Cyprus.

The confederation of southern Slavs in the uplands south of the Danube is breaking apart as the Croats, Slovenes, Bosnians, Hercegovinians and Macedonians, along with Magyars and Albanians, seek to shake off Serbian domination of Yugoslavia electorally. The federal army's advance on Zagreb halted in January 1991, and the Serb and Croat presidents began negotiations. The Slavs of the uplands north of the Danube have not resolved their differences, and there is a tension that could set Czech against Slovak. On the northern plain, the Poles are busy sorting themselves out economically, hoping that the Soviet Union to their east and Germany to their west will remain stable.

The greatest change in Europe's political geography was marked by the signing of the Four Plus Two treaty to end World War II in Moscow in September 1990. This recognition of a reunited Germany by the United States, the Soviet Union, the United Kingdom and France specified the Oder-Neisse line as the border with Poland. The united Germany, a member of NATO, is to have a maximum military of 370,000, and it pledges to have no atomic, chemical or biological weapons. The Soviets were bribed to withdraw their troops from the east. The historical record of what is now the most powerful nation in Europe makes many of its neighbors nervous. Before reunification there was pressure from the right in German politics to reassert the claim to Silesia and East Prussia, which are now in the hands of Poland and the Soviet Union. There have also been the first hints of a revival of the Nazi movement in eastern Berlin. The prospect of economic domination without the necessity for territorial occupation may satisfy the desire to extend the boundaries of the nation to the limits of *Deutschtum.*

Beside these momentous upheavals and dangers in Central and Eastern Europe, the ethnic and regional rubs in Western Europe seem to be of negligible significance. Although Flemings and Walloons, Welsh and Scottish nationalists, Irish republicans and unionists, Bretons, Basques, South Tiroleans and Corsicans have smoldered and sparked, these are not nation-wrenching scenes of mass slaughter. The economic stress between north and south in Italy, Spain, France, England, Germany and Norway finds political rather than warlike expression.

The Americas

Latin America. Across the Atlantic, the U.S. hegemonic hold on the Western Hemisphere, which reached its zenith in 1950, has been eroded by more than the defection of the socialist governments in Cuba and Nicaragua. The protection of U.S. investments and the support extended to the wealthy, land-owning descendants of the conquistadors have not maintained the upper hand everywhere. The size and sophistication of some parts of the Brazilian economy have established a degree of economic and political independence for its oligarchy. The wealthy ruling class merely keeps social turmoil and violence in check and protects their constituents' lives and property from the pent up frustration of the poverty of the majority, occasionally with violent severity. In Paraguay General Stoessner was overthrown in 1989, after ruling the country for thirty-four years with security forces who made their riches from smuggling. His successor is his son-in-law and former second-in-command, who promises to dismantle the police state. The Argentinian junta ceased to assist the Central Intelligence Agency (CIA) in trying to bring down the Sandinista government in Nicaragua in 1983, when the United States supported the United Kingdom in the Falklands war; Argentinian defeat in that year undermined repressive military rule. Politicians dependent upon election are less likely to act for the United States. Across the Andes in Chile, General Pinochet, whose regime owed its origins to CIA intervention in Chilean politics in 1973, was defeated electorally, and he stepped down from the presidency to the position of commander-in-chief. President Aylwin is attempting to circumvent the influence of the army and to create a representative polity, freer of political violence.

The greatest U.S. influence in Latin America now is the illicit demand for cocaine. The contest for the loyalty of the countryside on the eastern flanks of the Andes in Bolivia, Peru, Ecuador and Columbia is a fight for the loyalty and cash of the *cocaleros*. Coca growing is the most profitable use of this land. Those who protect the crop gain the support and revenues to rule. Coca is the main income and employment generator in Bolivia. The transmontane coca-producing area is where the Peruvian military and the violent extremists of Sendero Luminosa are currently fighting it out. The pathetic efforts of U.S. antidrug agencies to destroy the trade created by American affluence and moral malaise merely makes politics more complicated and violent in these nations. The illegality of the trade effectively endorses the power of the informal and landed sections of the economy in Columbia, for example, creating conditions conducive to the violent extraction of monopoly profits at crucial junctures in the production and distribution sequence. In Panama City they are still counting the cost of U.S. selective intervention, which caused much death and destruction in the campaign to remove President Noriega, whose involvement in the drug business had previously been cultivated and employed by the CIA. The members of the police force, with which Noriega ruled, were then given new uniforms and set back in charge. After a year U.S. troops are still in occupation, and democracy is still not in sight.

The outbreak of fighting in Nicaragua in November 1990 makes it plain what insurgency in Latin America is about. President Chamorro, a *latifundista*, had failed to deliver the small farms the Contras were promised when they agreed to surrender their weapons. Since they did not all hand in their weapons, their protests broke out into killing. The Sandinistas, who control the army, stood by the president who had ousted them from office electorally. In El Salvador the rebels who seek to overthrow the landowner's government are moving from guerrilla to full-blown conventional warfare in their bid for power. In November 1989 they took the center of San Salvador. They have turned to taking on the army in the open and have even shot down government aircraft. After a November offensive in 1990 the FMLN pressed for peace talks to end the eleven-year war. President Bush took the occasion of the shooting of a U.S. helicopter crew to restore $42.5 billion in military aid. The FMLN were contrite, but the fighting continues. The military still rule Guatemala with the *mano dura*. Concentrating on coverage rather than mobile fire power, they employ "extrajudicial killings" to protect the interests of the landholders and business elite from guerrillas who have been rebelling for thirty years. This is likely to continue under the new president who is a protegé of Rios-Montt and is beholden to the army.

The political fragmentation of the Caribbean is liable to sporadic violence like the Black Muslim outbreak in Trinidad in 1990. The U.S. excuse for intervention on geopolitical grounds, as in Grenada, has vanished with the retraction of Soviet global ambitions. In Haiti the election of Jean Bertrand Aristide, representing the poor of Port au Prince, was met with opposition from Duvalier supporters and the *tonton macoute*. The army stood by the new president, but violence continues.

North America. Apart from a fight about gambling and sovereignty in the Mohawk St. Regis reservation in April 1990, and the possibility of violence erupting over Quebec's status, the chief potential for mass violence in North America lies in the aimless misery of life for the black, urban poor. This dissatisfaction erupted in Watts, California, in 1967 and in Miami, Florida, in 1980, and so far nothing that would remove the cause for a future occasion has been done. If America is not socially or politically violent, the American population is incredibly violent personally. A cultural heritage that endorses the individual quest for retribution for actual or suspected wrongs, as well as the ready availability of guns, provides the setting for a death toll from murder and manslaughter that is greater than the killings of many wars of the tribal or national kind. The murder rate in the United States is four or five times that in Europe. The national capital tops the murder rate for U.S. cities, setting a new record every year. This is not interracial or cultural warfare; it is mostly confined within groups, with high rates in culturally homogeneous areas such as the Midwest. It seems to be a matter of the availability of guns. The British are more violent, but unarmed. People kill because they have a weapon at hand. Here is a people at war with themselves.

REFERENCES

War Data

Sivard, R. 1989. *World Military and Social Expenditures*. Washington, D.C.: World Priorities Inc.

Loss of Strength Gradient

Boulding, K. 1963. *Conflict and Defense*. New York: Harper and Row.

Theory of International Conflict

O'Sullivan, P. 1986. *Geopolitics*. London: Croom Helm.

Survey of Current Hot Spots and Potential Sites of War

Newspapers

Economist
Manchester Guardian Weekly
New York Times
Tallahassee Democrat

Book

Keegan, J., and A. Wheatcroft. 1986. *Zones of Conflict: An Atlas of Future Wars.* New York: Simon and Schuster.

5

THE GEOGRAPHY OF REVOLUTION

Chapter 4 put war into a geopolitical framework and addressed the competition for territorial influence among the rulers of nation-states. Although this competition has caused many of the wars that have punctuated modern history, others have arisen from internal efforts to overthrow the existing government. Our review of hot spots revealed a multiplicity of sources of discontent. To complete the picture of the geography of war, we need to explore the spatial dimension of revolutionary political change.

Neither prophets, nor practitioners, nor political theorists nor geographers have been very precise about the geography of revolution. Yet it is evident that geography plays a fundamental part in shaping the outcome of revolutionary violence. To remedy this, first we will examine the nature of political revolution and the significance of geopolitical position in determining its outcome. This leads naturally to an exploration of the geographical sources of revolution. Having discussed the nature and origins of revolution, we will have laid the foundation for an analysis of the geographical balance of power between revolutionaries and government. After this balance of power is described, a simple model of the spatial properties of this balance of force is presented in geometric form.

The intent in writing this is not to make the case that the outcome of violent competition for power is ultimately determined by environmental circumstances. This is not an exercise in geographic determinism. But it is important to point out that the setting, which is a matter of place as well as time, does channel the torrent of chance events that is history into structured outcomes. Location, situation and site advantages must be considered in the calculus of force, along with climate, terrain and the features of the human landscape.

RECEIVED WISDOM

What has been written on geography and revolution is inconsistent and imprecise. In 1847, in the *Communist Manifesto*, Marx and Engels remarked that the concentrated, urban masses of bourgeois-dominated society could be the setting for revolution. If we follow the evolution of socialist doctrine from this beginning, the favored venue for revolution has changed radically through a century and a half of political machinations. Lenin shifted the fountainhead of the force for change from Marx's Western European proletariat to the Russian proletariat in alliance with the peasants. A decade later Mao snatched up the banner of revolutionary leadership for the Chinese peasantry, and in the 1960s Lin Piao extended a share of the honor to the peasants of the Third World. The emphasis on guerrilla insurrection in the "countryside" of the world brought a stress on the exploitation of terrain. The memoirs and instructions of revolutionary warriors and theorists have a strong microgeographical flavor to them, emphasizing the value of attention to the details of the landscape. This has created a lore about geographically advantageous conditions for revolution, which does not stand up under close scrutiny. Although guerrillas may survive in wooded hill country, that is not necessarily where victory is won.

Academics have brought an array of sociological, psychological, economic and political theories to bear on the phenomenon of revolution. But this theorizing, be it functionalist, structuralist, behavioralist or utilitarian, is devoid of geographical content. Most theories are cast in terms of the time dimension. Revolutionary activity is deemed to display a temporal regularity. Revolution has been viewed as a predestined process in some overarching sequence of events, independent of the circumstances of place. The reality is that, as in any catastrophic event, the final outcome of violent conflict is a touch-and-go affair, wrought among the uniques of time and place. The manifold of possible futures, which is inherent in all of mankind's doings, explodes in the tempestuous swirl of uncertainty generated by political turbulence. The prospect of predicting the onset and outcome of revolution with law-like precision is remote. We are not certain of the causes, events and outcomes of past revolutions, never mind predicting the future. In particular, our grasp of the spatial dimensions of revolution is limited.

Radical geographers have concentrated their attention on the inconsistencies of capitalism and paid scant attention in their research to the violent events of revolution. The few excursions into the revolutionary landscape are mostly concerned with identifying places with the heroes and villains of revolutionary mythology. Some work has been done to correlate the Maoist doctrine governing the stages of revolution with the sequence of the geographical spread of insurgent control in China, Cuba, the Philippines, Greece, Java and Vietnam (McColl 1969). Perhaps the most valuable contribution to an understanding of the geography of political violence, however, is found in A. Wheatcroft's *The World Atlas of Revolutions*, which provides maps of the regions of discontent and the incidents that accompanied revolutionary change.

Here we examine the opportunities and inhibitions presented by geographic circumstances to the pursuit of radical political change. In particular a simple, geometric model of the territorial balance of power between government and insurgents is presented in the hope of clarifying at least some of the material issues involved.

THE NATURE OF POLITICAL REVOLUTION

To overturn the existing political order, it is necessary to overcome our enormous inertial preference for the way things are. We are creatures of habit. Most human behavior is a habitual repetition of what has been learned by imitation. Radical departures in social arrangement have been few, sometimes long drawn out, and far between until the last 300 years. Before this there were two chief modes of change. One was protracted competition between economies, involving the infiltration of territory by those whose way of life sustained a denser population. The other source of change was more sudden—a conquest by the militarily superior. This conquest was accompanied in some instances by slaughter and the replacement of population. Elsewhere it involved the imposition of a landlord class, or the substitution of an alien for a native elite. In the recent past, however, there have been a number of notable instances where the contest for the legal monopoly of violence that underpins the authority of government was internal.The outcome of these cases has involved not merely the replacement of the upper echelon of the ruling class—the kind of change signaled by the cry: "The king is dead! Long live the king!"—but rather fundamental, constitutional changes in government, accompanied by attempts to rearrange society according to some preconceived ideal.

This era started in England in the 1640s. The parliamentary victory, confirmed fifty years later with the flight of James Stuart, provided the precedent for the American Revolution of the 1770s, which in turn inspired the French Revolution of the 1790s. The radical departures of the French exercise then provided the model and slogans for an explosion of revolutionary fervor, which spread throughout Europe and around the globe through the nineteenth and twentieth centuries.

Despite the temptation to employ the imagery of fire or disease, even a cursory glance at the history of violent changes and the people involved over the last three centuries should convince us that this is not a contagious diffusion, spreading from neighbor to neighbor like a wave surging across the face of a nation, washing over its borders into contiguous lands. The domino-like ripple of revolutionary change across the face of the earth is a journalistic figment. Efforts made to construe violent political change as the result of a contagious process spreading from nation to nation have not been successful. The explanation of the geography of revolution lies not merely in the position of a nation with respect to other nations but, more important, in the internal conditions of its society and whether these present the goad and occasion for violent change. Certain changes in attitude, notions of what is right or wrong and information or misinformation about events may spread through time and space by contagious propagation and aid and abet those who would gain power by overthrowing the incumbent rulers. The geopolitical position of a nation may well create the opportunity for change, but the overthrow of existing rulers and the wholesale truncation and replacement of the governmental structure is, in the last analysis, a matter of the in situ chances for concerted action taken or missed by the various parties to a conflict. That the resolve of one party may be firm because of its ideology is possible, but for an elite cadre that vies for power, this fervor is not caught like a casually transmitted disease spread by contagious diffusion. The explanation for revolutionary zeal for change and power is usually more complicated and deeper seated than the occasion of a chance sharing of pathogens in air, water, body fluids or vermin. The revolutionary spirit obviously appears infectious to some. Wordsworth recalled being politically awakened by the French Revolution in 1789:

> Bliss was it that dawn to be alive
> But to be young was very heaven.

On the other hand, apart from fear of violence and loathing of change, there is a rational reaction against revolution, such as that of Edmund Burke, which is appalled by revolutionary abstraction and moral oversimplification.

The images of revolution as wildfire, pestilence or falling dominos are founded on fear, and they show a misconception of the nature of the transfer of political power. It is seized from above, not granted from below. The credit for a successful revolution can usually be given to incompetent and irresolute rulers. It is hard, given the strong preference among most people for the status quo, to lose the monopoly of force and the political initiative. How else could we explain the long persistence of institutions like slavery, landlordism and conscript armies?

Although fervor for change and violence may sweep through the countryside and provinces, revolution is crowned in the capital. Given that power is exerted via the manipulating of force and information, then the centers of the transport and communication networks must be held in order to rule. These are the geographical circumstances of revolution. Waves of passionate enthusiasm and insurrection in the provinces may keep the revolutionary spirit alive, but power must be grabbed where the geographical focus of society lies. The closest to revolution by popular acclaim and by the sheer mass of involved humanity was the victory of the mullahs over the shah in Teheran in 1979, when the street mob brought Khomeini to power.

The example of Iran brings to our attention the reality that revolution does not necessarily imply social progress, merely profound change. The motion imparted to society by revolution may be viewed as backward or forward, depending on the stance of the observer. Some turnovers have resulted in increasing political control over society, as in the case of the revolutions led by Lenin, Hitler, Mussolini and Mao. The outcomes of the Spanish civil war and Iranian revolution were antipolitical, but revolutionary nevertheless. Moral or ideological superiority is no guarantee of victory. The political orthodoxy which counts is that of the victors. Inasmuch as some revolutions have extended the circle of power to include more people, or increased the extent to which rulers consider popular response in making choices, they might be deemed to have enlarged some people's control over their destiny and, therefore, to have had morally beneficial outcomes. The historical evidence, however, leaves us in doubt as to whether revolutionary violence inevitably pays off with a heritage of greater liberty.

THE ORIGINS OF REVOLUTION

There is no wholly convincing theory of the social origins of revolution. Although Karl Marx's Hegelian historicism used to have a large following, its currency has been debased by events. The explanation for the political turmoil that translates into radical change in the personnel and style of government lies in the peculiarities of people, time and place. There are coexistent but diametrically opposed grand, teleological conceptions of history. One view is of an inevitable and benign historical process, punctuated by revolution, which governs the evolution of the social and political arrangements of mankind. In direct opposition to this view is a vision of revolution arising from a conspiracy of evil people to gain power over the world piece by piece in order to rule it in keeping with some diabolical doctrine. Neither of these notions is well founded in reality. These are religious positions, not scientific ones.

The fact is that successful revolutions have come from the right and from the left of the political spectrum. Whatever the myths of global dimensions used to bolster enthusiasm or antipathy, to attract support or to muster allies for a big power rivalry, the ultimate causes of success or failure of bids for power lie in the intricacies of local society, politics and geography as the setting for violent competition for power.

The historical record does suggest that the underlying prerequisite for attempted revolution is a regime that is considered vulnerable to overturn. It rather begs the question to say after the event that the necessary and sufficient condition for successful revolution is a government that can be overthrown. But, clearly, there must be a basic instability or fragility in a governmental structure that surrenders power. A structure that cannot withstand the loss of nerve or misjudgment of a particular ruler or official lacks the robustness required for evolutionary success.

A constitution and control structure with an abundance of apparently redundant links may survive the collapse of many elements in the chain of command, whereas a strictly hierarchical tree can be killed by cutting its trunk. The geographical manifestation of these differences finds expression in the size and disposition of towns and in the layout of transport and communication networks. A nation with several major cities only a little less important than the capital and with more fully connected road, rail, airline, telecommunication and utility networks is less vulnerable to any local damage to its body politic. Revolution may well have a greater chance of success in a nation with an overwhelmingly dominant capital, a strong radial component in its networks and constitutional arrangements and a minimum of circumferential linkages. This would explain why France has been more vulnerable than England to revolution, and it suggests that the United States is vastly less vulnerable to revolution than the Soviet Union.

It is not only the concrete structure of control that must be disabled in order for revolution to succeed. The mythical web that sustains the monopoly of force for a small elite in the face of overwhelming numbers—the source of sympathetic support, loyalty to a person or office or common identity—must be eroded or exploded. Clearly fear may stem the tide of such erosion. Swift, ruthless, concerted action to stop or delay the widespread expression of sympathy for change may counteract revolution. The savage reprisals of the British in India after the Sepoy Mutiny in 1859 kept a discontented subcontinent at bay for over eighty years. It seems to be the case that the degree to which a myth of identity between people and government is abided by has a geographical dimension. Clearly, states encompassing a considerable geographical variegation in language, religion, economy and wealth are less coherent. The notion of being Russian did not extend throughout the Russian Empire in 1917, whereas the notion of being American was fairly coextensive with the bounds of the United States by 1900. The invocation of patriotism was not enough to stave off revolution in Russia. It is noteworthy that the core of the revolutionary army consisted of Latvian riflemen. In America, fifty years before, enormous numbers were persuaded to flock to the Northern colors to preserve the integrity of the American nation in the face of the threat of radical change. On a more local scale, the more complete physical separation of Ireland, than Wales or Scotland, from England made the prospect of its revolutionary secession from the United Kingdom all the more possible. Irish identity could not be subsumed with the "British" tag as easily as the Scottish and Welsh identities.

On the revolutionary side, the other necessary ingredient of revolution is leadership, a person or group intent on overthrowing the present system and seizing power. The spontaneous upwelling of dissatisfaction will come to nought if someone does not channel restlessness into a force to divert the flow of history. There is a theory that revolutions occur after a period of improvement in the condition of life has ceased and a gap of dissatisfaction opens between still rising expectations and falling gratification. But, clearly, even if this were the universal source of discontent, this in itself is not sufficient to generate a violent bid to overthrow the government. Whatever its source, in order to manipulate discontent and focus it, the ambitious must develop or seize the means of communication or construct a system that is superior to the established channels of communication. The ultimate test of potency, however, is whether the revolutionary leadership can amass a greater potential for the exertion of force in the key localities of a nation than the government can. This talent is perchance violent, but not necessarily bloody. The Nazis seized power in Germany in 1933 with little bloodshed.

It is evident that the geography of revolutionary opportunity is simply the dual of the map of dangers to the survival of the existing order. Places with a negative value of safety for the government have a positive value for the revolutionary enterprise. Where the present rulers are weak, there is a chance for insurgents.

In order for revolutionaries to succeed, they must capture the places from which society is controlled economically and politically. Models or theories of revolution that imply a territorial evolution of the insurrectory state, starting with dissident bands in the mountains and forests and proceeding through stages of greater and greater geographical coverage, leading ultimately to the capture of the capital, merely emphasize the point that competitors for political power have not won until they seize the cities of the plains. The opportunities to do this have usually been created by external circumstances or impotence on the part of the incumbent, conditions quite unrelated in causality or historical logic to the aspirations of revolutionaries. They have just seized the moment. To then claim all of history as a conspiracy of events leading inevitably to their success is to credit revolution with a moral certainty which it does not deserve. Such a claim serves to provide legitimacy, just as spurious genealogies did in the days of kings. The reality of violent political conflict is a chance laden, circumstantial, geographical process with a heavy bias in favor of the survival of the existing order if it retains a firm hold on the nerve centers of society. There is no inevitable progress through mappable stages to the monopoly of territorial power.

THE GEOGRAPHICAL BALANCE

The fundamental significance of the spatial dimensions of revolution can be deduced from the simple observation that all political conflicts are territorial. They are contests for power over certain parts of the earth. Jurisdiction and sovereignty are defined in spatial terms. Violent changes in the established order have occurred where the pace and nature of change in society have left the rulers unsure and have undercut their monopoly of power and authority. Geographically, their power of command is thinned out to the point that it can be challenged openly with hope of success by dissidents.

First we will consider these matters on a global scale. Clearly, the circumstances of revolution are not confined within the boundaries of nations. The support of powerful allies is of enormous significance in determining the outcome of struggles for power, and the venues for such struggles can be either the outland or the cores of empires.

At the power core areas of the world, the speed and magnitude of demographic and economic change have sometimes outstripped the capacity of the established system of control so that the pressure for political change cannot be contained. This occurred in France at the end of the eighteenth century; these are the conditions addressed by Marx.

The balance of power at or near the imperial core area may also be upset by overextension of its resources of power by the government, reaching out over a wider sphere of influence and thinning its internal level of power. This becomes especially marked if it comes into competition with other hegemonic powers. These circumstances are found in the attempted revolution in Dublin in 1916 and in the Russian Revolution of 1917.

Revolutions have also succeeded on the thinning edges of the hegemonic spheres of the world and in the interstices between the proclaimed domain of great powers, where intervention is costly, extended over oceans or over empty land. Here, where the hegemonic powers either govern on the cheap or compete for clients, the opportunities for violent change are considerable. These circumstances occurred in the American Revolution and in those of Sun Yat Sen and of the communists in China.

From this broader, global scope we now focus down to the more local level of resolution, to explore the geography of the competition for power more closely.

If a pretender to power cannot amass sufficient force to take over the capital of a nation by overwhelming or subverting the government forces there, then the obvious path is to move toward the periphery, farther from the government's strategic reserve, buying time with space, giving the government the greatest opportunity to make a mistake. In any field or board game that plays out the rite of territorial competition, to attack a packed defense one adopts the indirect approach moving to the outside edges of the field. Here on the wings one picks up strength, strings out the opposition and gives them a chance to blunder. The objective is to draw out the defenders, creating space behind them to attack and get at the goal. Faced with the obvious defensive strategy of *cattenaccio* (forming a chain), revolutionaries survive or consolidate in the periphery. Evidently in the 1760s the thirty miles that separates Philadelphia from Valley Forge was enough.

Having offered a blood sacrifice in Dublin in 1916, the Irish republicans headed for the hills and turned to rural guerrilla warfare. The Long March from Kiangsi to Yenan was a similar move on a grander scale. Starting from a minuscule beginning, Castro headed for the woods of the Sierra Maestra straight away to pursue his revolution. This fugitive scurry was translated by Guevara into the positive doctrine of the *foco insurreccional.*

Che's transformation of a tactic of dire need into a necessary stage in the glorious unfolding of revolution is yet another example of idealization after the event of the military and political opportunities revolutionaries have seized or the stratagems forced upon them by circumstances. Revolutionary tactics and strategy have been sanctified with doctrinal significance, attributing to them moral force and, thus, historical legitimacy. The validity of this statement is all too clear when we look to revolutionary writings to find the geographical mainspring and sequence of revolution. Ideologists differ on where revolution should begin. Some say it started in industrial towns among the working classes, some among the poor and alienated of the big city and some in the countryside among the peasants.

Looking at the record of events reveals a variety of places where the pressure of change reached a critical level. The mob in the center of a big city provided the setting where dissatisfaction turned to violence in Boston in 1770; in Paris in 1830, 1848 and 1871; in St. Petersburg in 1917; and in Teheran in 1979. But, elsewhere, at other times, in France in 1797, in Vietnam in the 1950s and throughout the Third World since, it was conflict in the countryside that flared into revolt.

Turning to the geography of leadership, the land-owning or professional classes, with their greater leisure and mobility, their freedom from a particular locus and their driving ambition, have provided the organizers and directors of this violent energy in America, France, Ireland, Russia, China, Cuba and Angola. Probably the most effective seat of change has been in the profession of violence, the military. The sons of the gentry and wealthy and the officer class, with an essentially urban base but wide circumference of actual and potential connections, have the means of both fomenting and coordinating widespread local discontent. More often than not, intense discontent among those excluded from political power by the existing order but ambitious for their own interests reaches a critical threshold among the upper middle class of the biggest cities, which provides the leading cadres of revolution. These people spread the word, undermining existing authority among workers and farmers, first down the urban hierarchy from the capital along the national transport network and then by word-of-mouth diffusion from the terminals and junctions of the long-distance communication system toward the outer fringes and interstices of the nation.

When discontent turns to violence, it obviously has the best chance of catching and spreading in the dense setting of a city. Although a ferocious response to irritation may come more readily and easily in a bucolic setting, where the force of authority is lightly disposed, sporadic rural flare-ups have a thinner tissue of humanity on which to catch. There is less chance that the countryside will erupt into general conflagration and spread turmoil to the towns.

Once the conflict between government and revolutionaries is joined in the open, then the same geographical influences that come into play in any form of warfare apply. The offensive defense of the "little wars of the shadows" nags and picks at governmental power and may win a political victory; however, for unaided military success, revolutionaries must amass an army capable of beating what forces the government can command in order to seize the core of the nation. This has frequently involved converting large elements of the existing army to the revolutionary cause. Either this strategy, or the intervention of the conventional forces of a sympathetic power, holds the best hope of success.

GEOMETRIC REPRESENTATION

Revolutionary geography can be viewed in the abstract with the help of the notion of the gradient of power (Boulding 1963). We can conceive of power as being disposed continuously as a field decreasing regularly in magnitude with distance from a peak in the national capital. For simplicity's sake, we will pursue these thoughts in a rectangular spatial setting. All action and interaction take place along the lines of an infinitely dense square lattice of routes and links. Thus the perimeter of power will be square. If the rate of attenuation of power is a constant drop in power resources deployed per unit area with distance from the capital, then the contours of power will form a pyramid as in Figure 5.1.

The total volume of power, N, will be given by

$$N = \frac{(r \div 0.707)^2 \text{ x } h}{3}$$

where r is the radius of power and h is the density of power deployed in the capital O. If the government wishes to extend its radius of power, r, with the same total of resources, N, this can be achieved only by a reduction of the density of power at the center of the nation, h. If it wishes to intensify its concentration of power near the center (i.e., increase h), with a fixed total of resources, N, this must be bought with a retraction of the radius of influence, r. The nation is viewed as a square, even plain of radius R, which may be greater than, equal to or less than r. The borderlands of the nation may be unmanned by the apparatus of power, or power may extend beyond national borders. The institutions of nation-states and international affairs imply a sharp drop in the capacity to exert power within the frontiers of another state in peacetime, but one can image the extension of the radius

Figure 5.1
The Power Pyramid

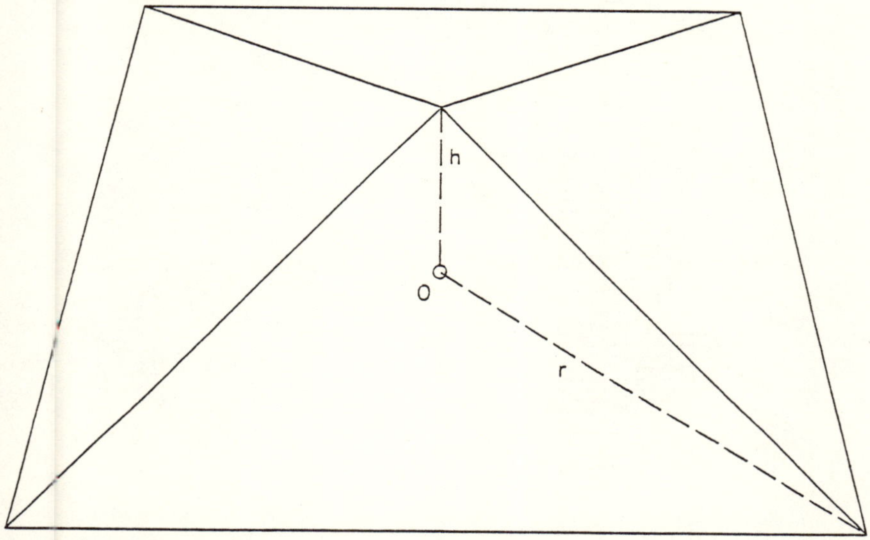

of power beyond a step-like drop at the borders into a neighboring state. In war, this inhibition vanishes.

This representation of the falling off of power with distance from the national core area is intended to reflect the effect of distance friction on transport and communications and, perhaps, a decline of concern with increasing distance from the core area. Beyond the national boundary, it can be taken to reflect not only the difficulty of deploying power physically, but also the loss of confidence of its deployers with distance from the safety of their home base.

Whatever its cause, if the power of government falls off in this fashion, then an attack on the status quo with limited resources can be mounted only at the periphery. Indeed, since government power may cease or drop sharply at the national border, for very small revolutionary forces a refuge beyond the frontier protected by a neighboring state hostile to the government may be the only possible venue. This was the strategy of the Algerian rebels and the Polisarios. But, the reach of political power is not limited by borders everywhere, and the South African regime reached out to strike at the ANC and its protectors with impunity, since it could mobilize an army twice as big and better equipped than the combined forces of its seven, black neighboring states, even supposing that they were to cooperate with one another.

If the rebel force is large enough to make a dent in the power field of the government, their dominance of territory will be determined by whichever

Figure 5.2
Peripheral and Internal Rebel Positions

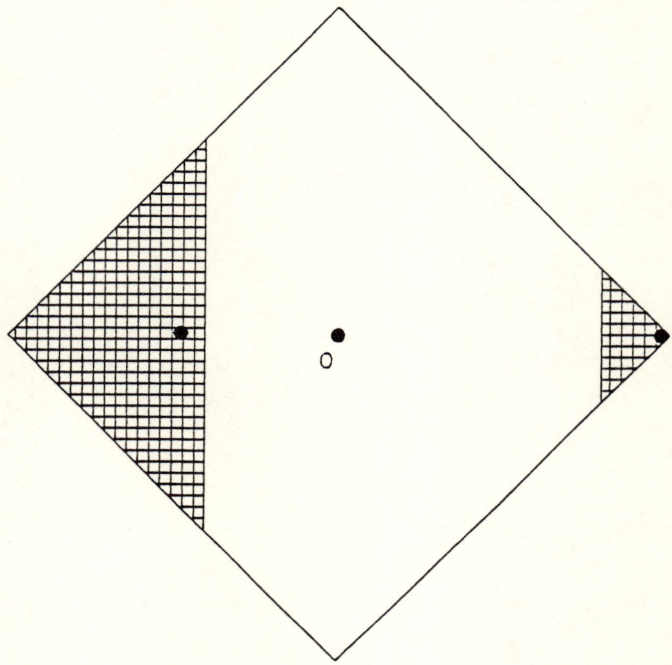

has the greatest power density at a point. The boundary between rebel and governmental territory is given by the projection of the intersection of their power pyramids onto the plain. Suppose the rebel force diminishes in power with distance from their command center in a fashion similar to that of the government, constrained, perhaps, by the same physical and psychic limitations. Then the revolutionaries will dominate the largest amount of territory by pushing the central axis of their pyramid of power as close as they can get to the capital without having their power peak overwhelmed by the government's gradient. Figure 5.2 shows the division of territory for peripheral and internal rebel positions if the rebel pyramid is ½*h* high and their radius of power is ½*r*.

If the revolutionaries' rate of decline of power with distance is greater than that of the government, they can penetrate nearer to the capital with the same total of power resources. The cost of this is a diminution of territorial coverage. There is a trade-off between threatening the capital more closely and holding more territory and hoping to amass more power from this in the future. We can illustrate the revolutionary options of sustaining more territory on the edge or threatening the capital more closely with lesser territorial coverage with an example where the height of the

Figure 5.3
Rebels with Steeper Power Gradients

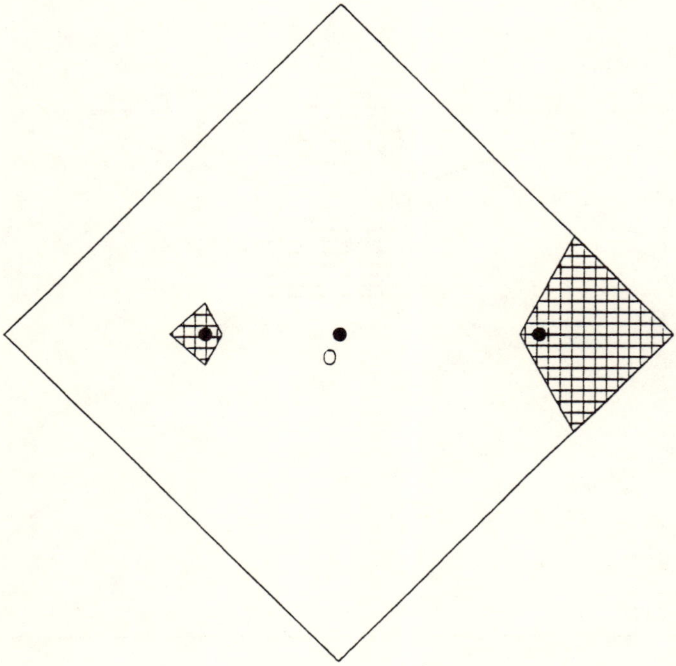

rebel pyramid is $\frac{2}{3}h$ and the rebel radius of power is $\frac{3}{8}r$, which gives a gradient twice as steep as that of the government (this, of course, requires a somewhat larger volume of power than the $\frac{1}{2}h$, $\frac{1}{2}r$ pyramid). The two positions are shown in Figure 5.3.

The revolutionary efforts considered so far could provide the basis for secessionary success in the peripheral provinces of an empire, for example, Ireland and Algeria; however, until the rebels are installed in the center, the imperial core will stand.

If the government cannot react swiftly in the deployment of its power, then there is a chance for a smaller revolutionary force to seize the capital. If the pinnacle of the revolutionary power pyramid stands higher than that of the government, the rebels may capture the center of the nation, leaving the government the periphery (see Figure 5.4).

If the revolutionary radius of power is only half that of the government, it can achieve equality with the central power density of the government; in other words, an equal value for h, with a total force only one quarter that of the government. Having seized the center, the revolutionaries are in a good position to hinder the retraction of governmental forces. Indeed,

Figure 5.4
Rebels Take the Center

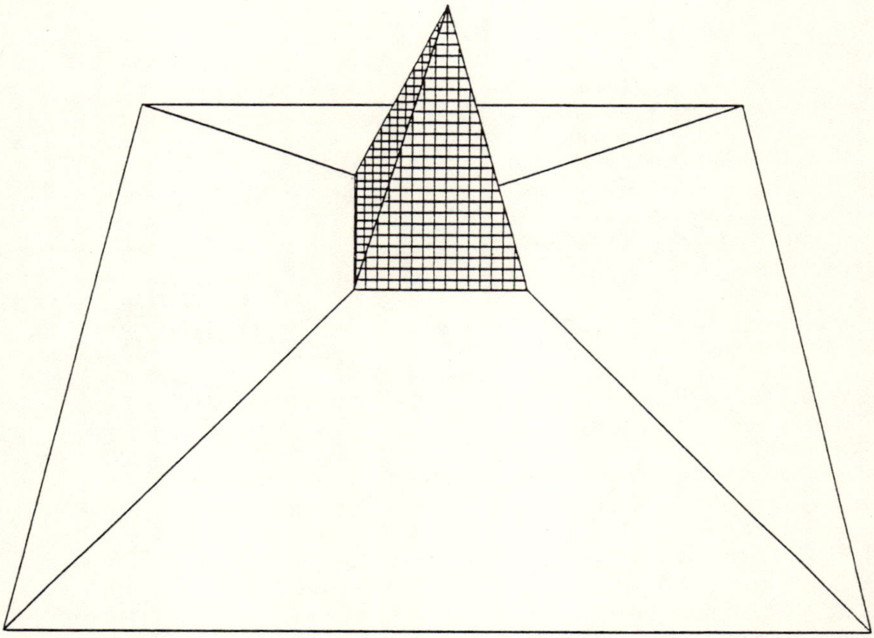

when one has the capital, the game is usually won. However, the communards of Paris learned in 1871 that this is not a rule without exception.

The other occasion of opportunity for revolution occurs when the government extends its radius of power to raise the power density at or beyond its boundaries in times of war. Even if more resources are employed for the exertion of power in wartime, its deployment will inevitably be extended in the direction of the enemy without, which is an opportunity for the enemy within. This state of affairs gave the Bolsheviks their opportunity in 1917.

The revolutionary prescription arising from this simple exercise would seem to be to keep one's forces concentrated within a tight radius of operation and to drive for the capital as soon as possible. Rebels can survive with much real estate at their command in remote and peripheral regions, but they cannot achieve sovereignty over a nation until they seize its core. For those in power, the lesson is to avoid being drawn into peripheral operations at a cost of thinning power in the core area. Since rebel forces usually have the advantage of greater mobility, there is the prospect that they can create space to attack the capital by drawing government forces out to the fringes and then filter through their lines to reconcentrate against lighter opposition. If the government can withstand the embarrassment of "no-go" areas for its writ within its boundaries, it can survive as long as it dominates the economic and demographic core area with the confidence of most of the people there. It seems that in Northern Ireland the British government took this stance with respect to the urban periphery in the Falls and Bogside of Belfast and Derry, as well as the rural periphery of south Armagh.

Reducing the question of revolution to a crude assessment of power, control and mobility of rebel and government forces represented by the height, breadth, slope and location of symmetrical pyramids may seem crass and minimalist. But reducing the treatment of these gross geographical factors to an abstract representation amenable to simple analysis clears the way for a concentration on other elements of the physical and human landscape, as well as the social and economic aspects of revolutionary change, with greater precision and clarity. It provides for, at least, a classification of geopolitical circumstances. The approach to greater understanding through the employment of strong cases, and what some regard as offensively simple-minded assumptions, may get us farther than intuition and anecdotes. Some of the simplification used here was, of course, for mere graphic convenience. What was presented here in simple linear geometry would hold for any monotonic power gradients and any continuous geometric space.

REFERENCES

Boulding, K. 1963. *Conflict and Defense*. New York: Harper and Row.

Davies, J. 1962. "Towards a Theory of Revolution." *American Sociological Review* 6(1): 5–19.

Harvey, D. 1979. "Monument and Myth." *Annals of the Association of American Geographers* 69(3): 362–81.

Huff, D. L., and J. M. Lutz. 1974. "The Contagion of Political Unrest in Independent Black Africa." *Economic Geography* 5(4): 352–67.

McColl, R. 1969. "The Insurgent State: Territorial Bases of Revolution." *Annals of the Association of American Geographers* 59(4).

O'Sullivan, P. 1982. "Antidomino." *Political Geography Quarterly* 1(1): 57–64.

———. 1983. "A Geographical Analysis of Guerrilla Warfare." *Political Geography Quarterly* 2(2): 139–50.

Rezendes, S., and P. O'Sullivan. 1988. "Terrain and the Effectiveness of Guerrillas." *Papers and Proceedings of Applied Geography Conferences* 9: 234–39.

Taylor, S. 1986. *Social Science and Revolutions*. New York: St. Martin's Press.

Wheatcroft, A. 1983. *The World Atlas of Revolutions*. New York: Simon and Schuster.

Wordsworth, W. 1850. *Prelude, or Growth of a Poet's Mind*. Book xi, l.108.

6

THE GEOGRAPHY OF BATTLES

The last two chapters considered the geography of the causes of war. The scale of this consideration was a geopolitical one, and the decisions evoked were in the realm of grand strategy. This sets the scene for the principal concern of this book, which is tactics and terrain. We must bring our attention down closer to earth and dwell on strategic and tactical choices and on the part that geography plays in these. The best introduction and source of data for generalizations along these lines is a survey of the geography of battles.

Constructing a classification of battles that will allow us to learn about the geographical circumstances that surround them presents a considerable problem. There are many dimensions along which to array the characteristics of violent conflicts. They can be arranged by historic eras; by whether they are fought on land, sea or in the air; whether they are fought in forest, desert, grassland, farmland, on mountains or on plains; according to the numbers and arms involved; or according to the nature of the political conflict involved. Because no one dimension captures the significance of geography to the outcome of battle adequately, various criteria are employed here to capture the evolutionary and timeless aspects of the relationship between fighting and landscape.

There is an obvious first cut to make between battles ancient and modern: The advent of rifled, breech-loading guns marks the start of the modern era around 1850. These guns extended the distance of engagement and, combined with railways, increased the scope of battle beyond the range of vision of one individual, so that generalship became a matter of strategy rather than tactics. Before this men killed each other in war at 100 yards or less, and commanders directed the movements of tight formations within their eyesight. In ancient battles, then, a general's success depended more on tactical ployes, employing the lie of the land that lay before him, rather than on deceitful stratagems, manipulating the forces beyond the horizon. We can trace the changes in defensive and offensive advantages that came with innovations in arms and the tactical responses to these by examining various battles from a series of historical eras.

The transition to the modern era is marked, in the first place, by a number of colonial battles between traditional warriors and modern armies, not always to the advantage of the armies. Modern warfare began with the American Civil War in the early 1860s and the Franco-Prussian War of 1870. This war and the Paris Commune provide an introduction to urban battles and their variety and changes over the last century. The last of these battles to be considered here, Panama City in 1989, involves an insertion of forces from a distant source. This leads us into battles arising from amphibious insertion and airborne insertion. While we are in the air, battles to dominate the air over a particular territory are worth a mention; however, our attention is focused on land battles for supremacy. The big struggles have taken place in either a desert or a middle latitude setting, and these are our final categories. Under each heading, only a few examples, chosen for the significance of the geographical factors involved, are examined.

MEDITERRANEAN STATES AND EMPIRES

The disciplined, heavily armored infantry of the Mediterranean peninsulas first showed their paces at Marathon in 490 B.C. On a small crescent of coastal plain on the north shore of Attica, six miles long and two miles wide, surrounded by limestone mountains clad in pines, olives, cedars and shrubs, Datis landed and deployed his Persians with a heavy superiority in cavalry over anything the Athenians could muster. The ill-drained horns of the plain were, however, turned to marsh by the rains of autumn, and the room for maneuver was limited. The Athenian commander, Miltiades, prepared his forces in the safety of the hills and employed the tactics of the refused center to outflank and crush the Persians in a setting where their cavalry superiority could not be brought to bear. The Athenian hoplites had trained to run in armor, and Miltiades sent his central phalanx running at the center of the Persian line to close before their cavalry could mount up. They struck the hastily formed Persian lines, then fell back drawing Datis' forces after them between the strength Miltiades had concentrated on his wings. These then smote and crushed the advancing Persians, killing 6,400 of them at a cost of 102 Athenian deaths. Hannibal also used the refused center successfully at Cannae in 216 B.C., only to have Scipio Africanus employ a variation on the theme against him at Ilipa, near where Seville now stands.

Returning to Greece, the series of battles fought at Thermopylae illustrate the significance of passes, the opportunity they present for a stern defense and the dangers of not covering alternative avenues of approach. The pinch in the coastal plain to a narrow passage between the mountains and sea at Thermopylae functioned as a choke point on the major overland route from the north to the core area of classical Greece in Boetia, Attica and the Peloponnesus. In 480 B.C. Leonidas and his Spartans held Xerxes' Persians at bay there until another route was found for the Persians to attack from the rear and slaughter them. In 279 B.C., the Greeks held the Celts and were then outflanked once more. In 191 B.C. the Roman Glabrio knew his history and geography better than the Seleucid king Antiochus, who repeated the error of not watching his rear. Although it was no great display of geographical genius, the defeat of the Spartans at Leuctra in 371 B.C. is worth a mention for the tactical innovation by Epaminandos, the Theban commander, of the refused left. This exploited the natural tendency for an advancing line of hoplites to fall behind on the left wing, as each man sought the shelter of the shield of the man to his right. To take advantage of this, the Thebans were arranged to receive the Spartans with the left end of their line pushed forward and with lots of men. They refused to engage the trailing Spartan left but, rather, concentrated on smashing the advanced right wing and rolling the Spartans up from there. At Arbela in 331 B.C. Alexander overthrew the Persians with another classic display of tactical skill. Darius had failed to secure the heights surrounding the battlefield he was preparing, which allowed Alexander to see what he had. Macedonian maneuverability avoided scything Persian chariots by opening ranks to let them through. Then, when a chink opened in the Persian lines, Alexander was ready to drive a cavalry wedge in and rip them up.

Ambush is the time-honored use of the lie of land. At Lake Trasimene in 318 B.C., Hannibal hid in the woods and caught the legions of Flaminius in a column of route. The Romans, struck from the side and slightly above, were smashed against the shore of the lake. Three hundred years later in the Teutobergerwald, Herman sucked Varus and his three legions into a series of ambushes in broken, wooded country and cut them to pieces.

HORSEMEN

The military and political domination of horsemen, which displaced the Mediterranean infantry, was best developed on flat, open terrain. This put a premium on seeking a bit of high ground for defenders on foot. In A.D. 457 on the Mauriac plain near Châlons, the Romans showed how infantry could withstand the cavalry superiority of the Huns given even a small eminence, this in a battlefield selected by Attila to favor his horsemen. At Hastings in 1066, William's Norman knights showed how well-established foot soldiers' defenses on high ground could be broken. A combination of high-trajectory volleys of arrows and a feigned retreat of a broken cavalry charge, which drew the Saxon housecarls from behind their shield wall, brought Norman rule to England.

We can see the end of heavy cavalry preeminence in a number of engagements in the fourteenth century. Two battles in Scotland and Switzerland illustrate how highland foot soldiers with pole weapons defeated knights by exploiting their home ground. At Bannockburn in 1314 Robert the Bruce arrayed his mass of Gaelic spearmen so that they had the protection of a bog on the one hand and a medieval mine field of pits dug along the only line of advance left for English cavalry, which reduced the horsemen's maneuverability to a minimum. The Swiss highlanders from Alpine cantons induced Austrian knights to come after them up the valley at Morgarten in 1315. On the confined valley floor, a phalanx of halberdiers had a distinct advantage.

Elsewhere archery showed its paces, as the flower of French chivalry attacked English plundering expeditions in western France, and the longbow fired from well-chosen ground prevailed. At Crecy in 1346 Edward III turned to fight his French pursuers with a stream and a marsh covering his right side and a boggy hollow to his front, where archers slaughtered wave after wave of French knights in the mud. At Poitiers in 1356 the Black Prince, weighed down with loot, had to turn and fight. He secured his retreat by getting his wagons across a river, and then he dug a ditch, during a truce, to reinforce a hedge to his front. Prepared for archery, the French advanced on foot, taking fewer casualties than they had at Crecy. The Black Prince's army held, and he decided on a desperate mounted counterattack at what turned out to be just the right time. Caught between the mounted English and an attack from the rear by their Basque allies, 2,500 knights, the king, and the dauphin were killed. At Agincourt in 1415, Henry V was forced to turn and fight before he could reach the safety of Calais. He deployed his archers where a strip of ploughland narrowed to a defensible frontage between two flanking woods. Again the French attacked on foot and were slaughtered in a setting where they were bogged down with little space to maneuver and bring their arms to bear. At Tannenberg five years before, the Teutonic Knights had paid the price of inflexibility and had come to grief at the hands of Poles and Lithuanians. The Poles thereby gained access to the sea, cutting East Prussia off from the rest of Germany. The German push to the east was shoved back to the Oder, and the seeds of continuing conflict were sown.

GUNS

By 1500 gunpowder had provided for the development of artillery and firearms. Muskets reinforced the pole weapons of infantry, and cavalry lightened up and adopted the pistol. At Breitenfeld in 1631 the inflexibility of the massed pikemen of the Catholic League was exposed by the clever, combined command of infantry, guns and cavalry by Gustavus Adolphus. Cavalry with pistols became the active element in tactics. The English revolutionary wars displayed this; Cromwell's mastery as the cavalry commander of the New Model Army emerged in Marston Moor in 1644 and Naseby in 1645. By the time William of Orange and James II faced off across the Boyne in 1690, artillery had become more important, and William used it to great effect to take a defended river crossing.

One of the greatest victories by a master of the combination of musketry, artillery and cavalry was won at Ramillies, near Liège, in 1706 by John Churchill. The opposing French army was deployed on a concave, four-mile-long ridge with a river and two villages to their right and marshes to their left. In taking advantage of this setting, however, they had overextended themselves and could not easily shift troops from wing to wing. Churchill attacked their outer wings, stringing out their center, where he then applied the winning blow.

Three years later at Poltava, Peter the Great demonstrated how to win by picking and preparing the ground and letting the opposition take the aggressive initiative. Charles XII of Sweden had made the mistake of pushing too far into Russia in the fall, and he had failed to achieve his objective of forging a link with the Cossacks in the Ukraine. Long lines lost him a supply convoy, and his army was decimated by the winter. He besieged Poltava, near Kiev, in May 1709. In June Peter arrived with a large army and took a position north of Poltava on a low ridge with forest to the left and right and a river behind him. Then, in the only approach left, he had ten redoubts built in a T formation. Charles' predilection for attack led the Swedish army into this avenue where the redoubts broke up their unity and tore at their innards while the main Russian forces repulsed their uphill attacks. This virtually eliminated the Swedish army and their hegemony in northern Europe, and it signalled the dawn of Russian ambitions.

In 1759 Wolfe gave a masterful demonstration of the indirect approach and geographical initiative in the extension of British and French fighting to North America. The British won the citadel of Quebec guarding the St. Lawrence and New France by a stealthy approach on an unexpected route. Four-thousand men crossed the river and climbed the Heights of Abraham in the dark with two cannons, to come on the fortress from the rear. Montcalm came out to do battle with 4,500 men, but no artillery. The British held their fire until the French were ten yards away. It was all over in fifteen minutes, and both Wolfe and Montcalm were dead.

The grand finale for tactical generalship, with one man directing a visible field of battle, came during the Napoleonic wars. Napoleon's strategic genius and aggressive tactics have been celebrated in the subsequent development of military doctrine. But the man who beat him, Wellington, showed an even keener appreciation for the opportunities of the landscape with his more defensive stance. At Busaco in Portugal in 1810, Wellington won time for his withdrawal to Lisbon and the protection of the defensive lines of Torres Vedras which he had built in secret. He selected a ridge north of

Coimbra, placing his men below the brow of the hill on the side away from the enemy, sheltered from artillery fire. The steepness of the hill, indeed, prevented the French from bringing artillery with them in the attack. Lieutenant Shrapnel's invention, which had been tried out first in the Guianas, was put to use to great effect in repulsing Massena's army. Having won time, Wellington then continued his retreat through Coimbra to his twenty-nine-mile-long defenses before the French cavalry could get around the mountains and threaten his rear.

This action is in complete contrast with the Napoleonic style on display at Borodino in 1812. Napoleon's 130,000 defeated Kutusov's 120,000 at a cost of 30,000 French and 44,000 Russian casualties. The Russians had avoided a real battle when the Grand Army advanced from Poland. The Russians regrouped at Smolensk and then fell back to a position astride the old and new roads to Moscow. Kutusov anchored his defense on Borodino and Utitsa, villages on the two roads, and based his deployment on four redoubts overlooking a steep-sided valley. Napoleon's marshals suggested an outflanking sweep around Utitsa to the south, but Napoleon had fallen into a state of strategic lethargy and had forgotten velocity in his equation of power with the product of mass and velocity. Instead, he mounted a frontal assault on the Russian right and center and battered away until night. Kutusov retired from the field under cover of dark, and a week later Napoleon marched into a deserted Moscow. Moscow burned down in September, leaving the Russian winter to do its job as Napoleon's army limped back toward Poland.

Napoleon's last chance came at Waterloo. Again Wellington was buying time, waiting for Blücher's Prussians in this case. He deployed his 68,000 on the back slope of a ridge, with a reserve to intercept any attempt at outflanking. Napoleon scornfully drove straight at the center of the British line. However, his artillery failed to do much damage to well-covered formations. Ney's cavalry were cut up by twenty infantry squares. Even the Imperial Guard could not prevail against the English, who kept themselves concealed in wheat until the last minute. When the Prussian cavalry arrived and attacked the French right flank, the French collapsed. The Old Guard formed a square, refused to surrender with one word, "Merde!" and were mown down.

The beginnings of a radical change in warfare had been visible before this in North America. The rifle showed its power in revolutionary battles. At Cowpens, on the North Carolina piedmont in 1781, Morgan disposed his men in three concealed layers on a hillside. His riflemen at the bottom of the hill fired twice at the advancing British, picking off officers and noncommissioned officers (NCOs), then retreated up hill to the main infantry line. Both then made an orderly withdrawal until they were concealed in a hollow below the crest of the hill where they turned to deliver a volley and a bayonet charge to the surprised British, who were destroyed. After a later war was over, outside New Orleans in 1814, Jackson's men showed the advantage of rifled weapons fired from well-prepared, defensive positions, opening up continuous rotating fire from 300 yards. The British had chosen the obvious and defensible approach and paid the price.

COLONIAL WARS

Much of the activity of armies and the evolution of tactics in the nineteenth-century transition to the use of breech-loading rifles took place in wars of conquest where imperial troops with modern arms were pitted against more traditionally armed and organized warriors, not always successfully at first. Since there is much to be learned from these colonial defeats, we will consider two classic examples: the Little Big Horn in 1876 and Isandalwana in 1879.

In the first example, a flying column of mounted imperial troops sought to strike the summer home of a nomadic and warlike indigenous population. Their commander, Custer, made the fundamental mistake of dispersing where it was not easy to recombine, in the face of excellent light cavalry. The areas between the rivers in the High Plains are flat to gently rolling, but in this semiarid setting the valley sides drop steeply to form forty foot breaks overlooking a flat floodplain, with braided streams and a scatter of cottonwoods. The U.S. cavalry were mounted riflemen. They did not employ the *arme blanche*, the saber, or indeed the lance, for a crashing charge. Their horsemanship was poor. When they engaged, they dismounted and formed a skirmish line with every fourth man holding the horses, which reduced their firepower by a quarter. They had a good weapon, the '73 Springfield breechloader, but they were poor shots, not having had enough ammunition for target practice. The evidence suggests that they did not sell their lives very dearly; they killed very few Indians. What finally defeated the Sioux later at Wounded Knee was what Custer had left behind: the rest of the 7th Cavalry and the Hotchkiss guns.

Isandalwana was essentially an infantry engagement. A British expeditionary force intruded into the homeland of the Zulu, seeking a decisive engagement to destroy Zulu military power. The chief British weapon was another breechloader, the Martini-Henry. These weapons were faced with the stabbing spears and shields of the Zulu impis. Zulu military might was founded by Shaka in the 1820s. Traditional war among the Bantu was mostly display to achieve subjugation, not slaughter. The opponents drew up a spear's throw apart, threw a few spears and engaged in some individual fights. Shaka changed this. He toughened his men so that they could run barefoot for forty miles and then fight a battle. Their spear blades were broadened and shafts shortened, so they could use them as stabbing weapons at close quarters. Their shields were enlarged. The regiment, or impi, closed on any opponent rapidly; the fastest formed two horns to draw the enemy in on the head and loins in a double envelopment. With these weapons and tactics, combined with a ruthlessness that left no enemy behind, Shaka's *mfecane*, or crushing, consolidated an empire bigger than Napoleon's, driving others, such as the Xhosa and Ndebele, from Natal in a wave of dispersion. Shaka's successor, Cetewayo, faced with British opposition, employed these tactics on a strategic scale. He dispatched his impis to get behind the British, using their superior mobility over rugged terrain. Steep-sided valleys of intermittent rivers, again the features of semiaridity called *dongas* in this case, were used along with the tall grass of the veld to close fast to stabbing range, despite the Martini-Henry. Lord Chelmsford obliged Cetewayo by dividing his force and failing to take defensive precautions. The effectiveness of the rifle was nullified by screwed-down ammunition boxes without tools to open them. A few hours later, at Rourke's Drift, a clear field of fire and a regular supply of ammunition held the Zulu. Finally, in July 1879, the impis were broken at Ulundi by Gatling guns and artillery.

RAILWAYS AND RIFLING

The railways ushered in the era of national mobilization of enormous armies. Rifling extended the range of engagement to 500 yards for infantry and miles for artillery. The tactical implications of these changes became apparent in the American Civil War, although the lessons were not always learned. The logistics of supporting and moving hundreds of thousands of men came to weigh more heavily on commanders. U. S. Grant was the first to command an army of a million. Rifles and long-range artillery gave the advantage to the defensive. All the components of the entrenched, barbed-wire standoff of World War I were to be seen in the lines outside St. Petersburg in the Appomattox Valley in 1863. The deadliness of modern weapons was to be demonstrated with special clarity when war broke down into soldiers' battles in closed terrain. Among the woods along the Chickamauga Creek in Tennessee, Bragg and Rosecrans each lost 40 percent of their foot soldiers. When Bragg's men got the upper hand, he could not exert enough control to pursue and destroy the army of the Tennessee.

Elsewhere, however, the Civil War inspired the doctrine of mobility and indirectness which was to dominate thinking after World War II. Grant's victory at Vicksburg, when he captured the railway junction at Jackson as a strategic barrage to cut the Confederate line of communication eastward, was a masterful display of a war of maneuver. Sherman's extension of this western style of mobile war to Georgia probably won the war for the Union. The dependence of massive conscript armies on railways for transport and supply reduced the fluidity of war, giving a limited range of operations from the lines for rail-fed armies. These lines posed a problem of vulnerability and became targets for raiders such as Forrest and Morgan. Sherman, aware of the fragility of fixed supply lines, deliberately cut his army's reliance on rail-borne support and struck at the Confederates' rail dependency. He avoided decisive battle and sought to seize the lines and junctions of the Southern rail system, starting with Atlanta. The march to the sea, which involved five columns, threatened many possible targets and maximized Southern uncertainty. Changing direction at will, Sherman cut Georgia supplies off from Richmond, and then, to cut off overseas sources, he rolled up the ports of the Atlantic seaboard. Clearly, in this campaign the concept of battle had taken on a broader, more prolonged sense.

The battles that mark the beginning of modern warfare in Europe took place in Alsace and Lorraine in the early, decisive stages of the Franco-Prussian War of 1870. Here von Moltke directed a Clausewitzian drive to destroy the enemy's forces in battle, supporting this with careful railway timetabling. This met with success in the face of French strategic ineptitude and their lack of experience in large-scale operations. French generals had earned their spurs in the colonial wars.

Northeastern France presents a corridor between the Hunsrück and Ardennes to the north and the Vosges to the south, permitting passage from the Rhine Rift Valley to the Paris Basin. Armies have marched between northern France and central Germany time and again through history. The region consists of a series of elongated, north-south aligned plateaus with sharp, east-facing escarpments, separated by broad vales. The lower areas are situated at 300 feet above sea level; the rolling plateau surfaces are at 600 feet. The escarpment edges are fairly sharp rises of 300 feet, cut by gullies and ravines, often with outliers separated from the plateau massif. About one-third of the surface is wooded in the form of small, commercial woodlots free of any undergrowth with well-defined boundaries between woodland and farmland. The war, as was traditional, began after the harvest so that the fields were in grass or stubble. The settlements of the region are small villages of substantial stone-built houses. This, then, was where the war was won and lost.

The initial lines were drawn up from the Vosges to the Hunsrück, between the Rhine and the Moselle. In arms the French had the central pin-fired Chassepôt rifle; the Germans, breech-loading, rifled artillery. The key difference lay in the superior organization of the German armies and their preparation to fight a war of large-scale maneuvers. The French were arranged in divisions and corps, but had no experience of operating as field armies. Their greater articulation enabled the Germans to amass and overwhelm the French with superior numbers wherever they met. The French were also caught on the strategic hop. They had initially planned to defend on the Rhine at Strasbourg and on the Moselle at Metz, while organizing their reserves at Châlons on the Marne. Napoleon III interfered and changed the plan to an immediate invasion of Baden-Württemberg and Bavaria, hoping to induce the Austrians to join the French. However, the original defensive mobilization plans were still in effect when war was declared, and the French concentrated at Strasbourg under MacMahon and at Metz under Bazaine. Superior logistics enabled three German field armies to form along the northern frontier between the Rhine and the Moselle and to march south on August 2, beating the intended French crossing of the Rhine to the draw. Under the direction of von Moltke, the chief of the Prussian general staff, the First and Third Armies, on the west and east flanks, headed up the Moselle and the Rhine, respectively, while the mass of power of the central Second Army sought to grip the French by the throat. The Third Army's advance met with little resistance. Prussian infantry took advantage of wooded cover to shoot up the French cavalry. MacMahon retreated into Strasbourg, which the Germans tied up with one division and then marched unopposed to Nancy. MacMahon's army filtered through the Vosges and headed for Châlons. The German First and Second Armies were faced with French defenses on the escarpments and had to make bloody frontal assaults to advance. In the face of these, the French fell back on Metz, fearing for their flanks. The Germans swept around Metz, cut off the French line of retreat to Verdun and established themselves on the escarpment of the Côte de Moselle overlooking Metz from the west. They used woodlots to screen their movements and drove the French out of their wooded cover with artillery fire. Bazaine's army was pushed back into the fortress of Metz by August 19 and was pinned down there for the duration. The Second Army then joined the Third Army in chasing MacMahon toward Verdun. Bottling MacMahon and Napoleon in that fortress, the Germans marched on to lay siege to Paris. The battles in Alsace-Lorraine had been primarily infantry fights. The rifle accounted for 90 percent of the casualties. The rifle and broken topography negated the effectiveness of cavalry. There was still, however, room for maneuver. Armies were small and compact. Flanks could still be turned and key terrain seized.

CITY BATTLES

The German siege of Paris in 1870 provided the occasion for the first significant example of modern urban warfare. The military inclination in modern city design was to make the hearts of capitals indefensible in order to prevent the mob from seizing and defending the center of the nation. Baron Haussmann had replaced the medieval maze of central Paris with streets too broad to barricade and with long sight lines for artillery. The Commune of 1871 was put down in short order. The Army of Versailles easily outflanked any barricades the communards threw up. This pioneering venture in proletarian revolution failed because its leaders did not appreciate the change of technique necessary to fight in the new Paris against more lethal guns. The tactics of 1848 would no longer do.

What Lenin called the first red revolution took place in the streets of Dublin in Easter 1916. This was, however, more in the way of a gesture, a blood sacrifice to stir the nation to revolt. One of its leaders, James Connolly, who commanded the union-based Citizen's Army under the banner of the Starry Plough, had written in 1915 about the similarity between urban landscape and broken terrain. He foresaw that guerrillas would have an advantage in the townscape's equivalents of glens and passes among the hills. Sean O'Casey urged a guerrilla campaign. But, when it came to the action, the Republicans sought to be soldiers rather than guerrillas. They seized the open prospects of central, Georgian Dublin, the straight roads, squares and big public buildings such as the General Post Office, the Four Courts and South Dublin Union and Boland's Mill, strung out along the River Liffey. They occupied architectural targets and threw up barricades. Artillery, including the guns of the HMS *Helga* on the Liffey, blew them away. There was no widespread popular rising. However, when General Maxwell had the leading rebels summarily shot, with Connolly so badly wounded he had to be tied to a chair, nationalist, republican sentiment spread like wildfire, and the IRA took to the hills.

Early the next year demonstrations and strikes were held in Petrograd in response to Bolshevik demands for bread, land and peace. These demonstrations were scattered with the new equivalent of Cossacks: armored cars with machine guns. By November, the Soviet organization had recruited many of the sailors and soldiers in the capital to the revolt. The provisional government's forces gave way as the Soviets took over the city, with only a token fight in the Winter Palace. Lenin selected a regiment of Latvian riflemen, who were delighted to shoot any Russians he deemed it necessary to eradicate, as the core of the Red Army, the instrument of revolutionary power.

That cities can be defended if there is a will was demonstrated a quarter of a century later on the banks of the Volga at Stalingrad. The *Wehrmacht*'s 6th Army under Paulus was given the symbolic task of taking this city of little strategic significance. The army penetrated rapidly from the edge of town through the residential areas. Despite the attentions of Russian storm groups they got to the downtown within a week and took it after two more weeks of fighting. It was the industrial area along the river that was the sticking point. Big factory buildings proved to the advantage of the defense. Dust, smoke and close-quarter fighting eclipsed the effectiveness of the *Luftwaffe*'s command of the air. The Russians held out while winter closed in, along with a million Soviet soldiers. Surrounded, Paulus was forced to surrender.

The battle of Algiers in 1956–57 was a very different form of urban warfare in a very different type of city. Late in 1956 Generals LaCoste, Salan and Massu met to devise a way to deal with the fighting between European colonists and Muslims. They determined to concentrate on rooting out the Front de Liberation Nationale (FLN), the rebel organization. Under the leadership of Saadi Yacef, they were opening the gap between colonists and Muslims with a campaign of assassination and bombing. The 1,400 fedayeen of the FLN operated from the Casbah, the old Arab quarter of the city. The French mounted a campaign of aggressive patrolling and house-to-house searches. The structure of the Casbah, consisting of a number of communal enclaves with limited access between them, made isolation and control manageable with relatively few checkpoints. Using "loyal" Muslims to infiltrate the FLN, the French swept up suspects and used torture to collect information. Within three months, the major elements of the FLN had fled the city. Yacef countered with a bombing campaign aimed at colonists, but the French captured him and killed his hard core. Militarily, then, this was a French victory, but their ruthlessness alienated moderates. Yacef's goal was achieved. No political compromise was possible between colonists and Muslims.

To complete the picture of urban warfare we have a recent example of the insertion of a massive conventional force by a foreign power to topple a government it does not like. The U.S. operation against Noriega in Panama City in 1989 showed how easy it is to seize a city if one is not too sensitive to civilian casualties and has an overwhelming superiority in people and firepower close at hand. The Americans could amass most of their force in their bases in the Canal Zone. But since the same paramilitary police force, albeit in different uniforms, along with a continuing presence of U.S. troops, remain the instruments of authority, the nature of the change brought about is hard to discern.

AMPHIBIOUS OPERATIONS

The most significant insertions of military power from afar have taken the form of landings from the sea, involving the combination of sea, land and air forces. They require secrecy, deceit, the right weather, panache and luck for their successful execution. Failures in stealth, boldness and promptness turned what might have been a brilliant success for combined operations in Gallipoli in 1915 into a bloody mess which cost half a million lives. In the Atlantic theater in World War II the landings in Dieppe, Algeria, Sicily and Anzio prepared the way for the greatest amphibious operation ever in 1944 in Normandy. In the Pacific, Guadalcanal in the Solomons in 1942 was the first step that led to the destruction of Japanese military power in Okinawa in 1945, where the U.S. Marines elaborated the techniques of assaulting the beaches.

Since 1945, with the withdrawal from formal, imperial occupation of territory by powerful states, much of their interference in other's affairs has involved landing troops against resistance, and so the lessons continue to be learned. At Inchon, Seoul's port, in 1950, MacArthur gave a display of geographic audacity. To break out from the Pusan Perimeter, he set aside logic and invoked intuition, seeking to get behind the enemy where the difficulties were so great that the operation seemed entirely unlikely and, thus, came as a complete surprise. A landing at Inchon involved every possible natural and geographical handicap. All of the chancy circumstances involved fell to MacArthur's favor in the event, and the foothold was secured with a loss of only twenty marines.

An object lesson in how not to conduct a landing was provided by the CIA's Cuban army in the Bay of Pigs in 1961. It involved a serious underestimation of their opponents. Such an operation against odds of 14:1 has little chance of success. A lack of air support does not improve the odds. Neither does trying to establish a base of operations in a swamp. British success in the Falklands in 1982 can largely be credited to a half-hearted Argentinian defense. They failed to defend the obvious landing place at San Carlos adequately, and they failed to hold onto the heights overlooking Port Stanley. This was a demonstration that there are conditions around the globe where tanks and helicopters are not much use, and success in battle comes down to having soldiers fit and trained well enough to take advantage of the difficulties of the environment. But, then, we can ask does it matter. The final solution is likely to be a political one that might have been achieved without bloodshed if there had been a modicum of goodwill and a lack of ambition for glory, both by Galtieri and Thatcher. The very next year, Reagan put in the U.S. claim to military prowess in Grenada, with or without the permission of the sovereign power, that of Her Britannic Majesty. The United States had full command of the seas and air and an advantage of 10:1 over 679 Cubans, 629 of whom were construction workers. If one is going in for amphibious operations, however, these are the terms on which to do it. The glory was shared around in the form of 26,000 campaign medals for an operation in which there were only 7,000 combat troops. The gilt was a little tarnished by the snafus involved. Four navy SEALs, weighed down with equipment, were drowned in the preparatory stages. A SEAL unit trying to knock out Radio Grenada attacked the wrong building. Two Delta Force missions ended in near disaster. Eighteen helicopters out of eighty-eight were put out of action. Communications were a mess; the four services used different frequencies. One soldier, finding himself surrounded, used his AT&T credit card to call Fort Bragg in North Carolina, from whence the call for helicopter support was routed back to Grenada through the U.S. Navy. Americans were killed by their own fire. Marines used old British topographic maps; the Army used tourist maps. Because of faulty maps, a marine airstrike hit an army command post. Despite the wealth of experience and massive planning organizations of the military, things can go wrong.

AIRBORNE BATTLES

The history of airborne insertion goes back to the German invasion of Crete in 1941. The largest airborne operation ever mounted was Market

Garden which led to the battle at Arnhem in 1944. Montgomery mounted a combined airborne and armored thrust to the Rhine, capturing canal and river crossing with airdrops to prepare for the advance of ground forces. The Germans halted the armor's spearhead at Nijmegen on the Waal. This left the lightly armed British First Airborne Division a bridge too far ahead and faced by an SS Panzer Corps, air photo evidence of which had been ignored. The 2nd Battalion of the Parachute Regiment did take the bridge across the Rhine at Arnhem, but the Panzer Corps efficiently surrounded and eradicated them.

Helicopter insertion came into its own in the 1960s. Ia Drang in 1965 was the first full-blown engagement in which helicopter-borne and supported U.S. troops were pitted against the North Vietnamese Army (NVA). The setting was a heavily dissected landscape with dense forest on slopes and in valleys, with grassy hilltops. Because of the heavy forest and tall grass, the forces involved came to grips at close quarters. The battle degenerated into a soldiers' war, with loss of control and severe limitations on supporting fire. The helicopter provided mobility to U.S. forces, but triple canopy forest greatly reduced their capability for reconnaissance and fire support. Casualties were high on both sides. This experience provided U.S. forces with a first taste of the limitations placed on firepower-based tactics in dense vegetation and a glimpse of the potential of the helicopter in a forested setting, along with the inversion of terrain this creates. The offense takes the high, open ground and attacks downhill into the defenses in the woods.

Another innovation of World War II was Orde Wingate's stronghold strategy, applied first in Burma. This consisted of an air-supplied garrison placed in a remote area where the enemy held sway. General Salan translated this into counterinsurgency terms in 1953 at Dien Bien Phu. The object was partly flag-showing, but it was also intended to draw guerrillas out into the open, irritated by a center of resistance in their midst, where the superior firepower of the imperial forces would tell. The *base aero-terrestres* was the established order's version of the *foco insurreccional*. However, if the rebels can muster enough firepower of their own, as Giap did, and if a base is poorly located in a depression with wooded firing platforms all around it, this strategy is unlikely to succeed. Khe Sanh in 1968 was in some ways a miniature replay of Dien Bien Phu, with Giap now dealing with the U.S. Marines instead of the French Foreign Legion. Khe Sanh is a village on Route 9 from Quang Tri to the Laotian border. In 1967 the Marines established an air-supplied base nearby, from which to strike at NVA infiltration routes through the demilitarized zone. The NVA bypassed its reach easily and it lost its military function early on, but it took on a symbolic significance. Westmoreland vested it with importance and vaunted impregnability, claiming that it was a honey pot attracting the NVA so that B52s could smear them. It is puzzling that its vulnerable water supply was never cut off and that enough planes got through to maintain food supplies. There is a suspicion that Giap reversed the honey pot rôle, leaving it standing to soak up men and firepower while he prepared for the Tet offensive.

Since it does illustrate some of the connections among the geography of the theater, strategic objectives, technology, tactics and chance, it is worth dwelling for a moment on one entirely airborne engagement, the Battle of Britain, which was fought in the late summer and autumn of 1940. The Royal Air Force (RAF) and the *Luftwaffe* fought for supremacy over the English Channel—essential for a German invasion, or its British repulse—along a 500-mile frontage of southeast England. Goering first sought to entice British fighters into doing battle with superior German numbers by targeting selected British locations. Radar and an efficient ground control system allowed the RAF to concentrate more Hurricanes and Spitfires than Messerschmitt 109s, which they could outfly anyway. Goering's reaction to this was to switch to attacking airfields with bombers. This was succeeding when chance and fury came to Britain's aid. At the end of August, a *Luftwaffe* bomber strayed over London and dumped its load. Churchill ordered an immediate reprisal on Berlin. Blinded with rage, Hitler ordered that all attention be focused on London. This took the heat off Fighter Command and made their task easier. They now had a small, clearly defined patch to defend. In September 1,000 German bombers attacked London while RAF Bomber Command wrecked the German invasion barges lined up in mainland ports, ending that ambition. *Luftwaffe* losses from daylight bombing mounted, while Beaverbrook replaced RAF losses at an amazing rate. Finally, in October, Goering switched to nighttime bombing. The Battle of Britain was over, and the Blitz had begun.

DESERT BATTLES

If the Battle of Britain was the first stopping point in British fortunes in World War II, the turning point was fought in the western desert of Egypt and Libya in 1942. Montgomery with 150,000 men and 1,114 tanks faced Rommel with 96,000 men and 600 tanks along a strip of desert between the sea and the impassable softness of the Qattarah Depression, so that outflanking sweeps were impossible. The British had command of the skies. The British attacked, and after a week of blocked thrusts and feints, with the Axis forces out of fuel and down to 35 tanks, New Zealanders broke through the German lines on November 2. Despite Hitler's orders to hold the line at all costs, Rommel extricated what was left of the Afrika Korps and escaped, leaving the Italians to their fate. Montgomery failed to pursue, but El Alamein had turned the tide. This was the end of the beginning.

The battle at Mitla Pass illustrates where desert warfare had gotten to by 1956. The occasion was the Suez crisis. The Israelis had agreed to attack Egyptian positions in the Sinai Peninsula, posing as a threat to the security of the Suez Canal to provide Britain and France with an excuse for seizing it. The Sinai is an unproductive triangle of land consisting of a broad strip of sandy desert along the Mediterranean coast; a central strip of dissected plateau crossed by tracks following dry wadis, such as the Mitla Pass; and the Sinai Mountains forming the southern portion of the triangle, with the base of Sharm al Sheikh at its tip. This provided the opportunity for Sharon to demonstrate the armor-led driving advance that was to become the Israeli specialty. It advanced 190 miles in twenty-eight hours. What was also displayed was its weakness, the tendency to get strung out and for the spearhead to lose contact with the shaft. The Egyptians turned Sharon back in the narrow defile of Mitla, and the Israelis had to cut across the desert, avoiding the pass, to get to the town of Suez and the road along the west coast to Sharm al Sheikh. The Six Day War in 1967 was a replay. The Israelis established the upper hand with a preemptive attack on the Egyptian air force on the ground. Armored tactics had evolved; the Israeli *ugdah* faced the Egyptian copy of the Red Army's sword-and-shield defense. Despite a monopoly in the air and Centurion tanks, the Israeli advance to the canal was not as rapid as expected and was not without loss, especially where the Egyptians regrouped after the armor had passed, leaving the infantry and artillery in the lurch. Air superiority told, however, and after a start on June 6, the leading Israeli elements got the canal by midnight on June 9.

For the Yom Kippur War in 1973, the Egyptians had built up forces and preparations in secret to cross the Suez Canal and assault Israeli defenses on the east bank. They surprised the Israelis enough to establish bridgeheads, and hard fighting ensued in what was now the Israeli buffer zone of the Sinai. The coordinated Syrian attack in the north struck at Israeli defenses between Mount Hermon and the Jordanian border. The Israelis held, turned and drove the Syrians from the high ground, showing the limitations of the tank in difficult terrain faced with prepared defenses.

Operation Desert Storm, in which the U.S.-led coalition forces drove the Iraqi army from their dug-in positions in Kuwait, demonstrated once more the power of unopposed command of the air and the lethal efficiency of planes and missiles in well-known terrain with little cover and clutter. The Iraqis were clearly too demoralized by the bombardment they had sustained to make much of a fight of it, and their retreat turned into a rout.

BATTLES IN THE MIDLATITUDES

The turning battle of World War II was fought at Kursk in the Ukraine in 1943. This was Hitler's effort to regain the initiative on his eastern front by pinching off a bulge in the Soviet line around Kursk. What ensued was the biggest tank battle ever fought with two million men, 6,000 tanks and 4,000 aircraft. The Soviets knew what was coming, and they held the line and countered. The German withdrawal turned into a rout, especially after Hitler pulled out troops to counter the Allied invasion of Sicily. In the fighting, *Luftwaffe* superiority in the air was occluded by summer dust and smoke, and the greater speed of the Soviet tanks outstripped the better guns of the *Wehrmacht*'s tanks. In the Ardennes, in the winter of 1944–45, the Battle of the Bulge was Hitler's last offensive on his western front. Ten armored divisions and fourteen others drove along the grain of the Ardennes westward hitting the American sector of the Allied front where it was lightly manned, reaching for Antwerp, trying to split the Allies as they stood ready to attack across the Rhine. The *Wehrmacht* advanced rapidly through the snow, protected by a heavy cloud cover, and reached Dinant on the Meuse. But Allied resistance stiffened, the skies cleared, German fuel ran out and their drive lost steam. Faced with counterattacks, Hitler pulled out the Panzers to meet the expected Soviet winter offensive in the east, and the bulge collapsed. However, losses of men, aircraft and tanks had been too great, and they did not have enough left to resist the Red Army's advance to Berlin.

It is fitting to end this historical survey with a reminder that there are temperate plains outside Europe and that violent turning points in history have been fought on these. Xuzhou was the crucial battle of the Chinese Civil War, one in which there were a million people in action on each side. In 1948 the Nationalists made a stand at this juncture where the railway from Beijing to Nanjing crosses the line from the coast to the middle reaches of the Huang He. The Grand Canal from Shanghai to Beijing also passes in this vicinity. They might have done better to fall back farther south to the Haui He. The river and a string of lakes would have provided a natural defensive line. Xuzhou was defensible, but the surrounding countryside was open and featureless, with no inhibitions to wide, encircling movements. The Nationalists made the mistake of withdrawing into the city from a wider perimeter. This disarmed the effectiveness of the American equipment with which they had been provided. It also provided the Communists with a clear field to sweep around them and press them from all sides. Liu, the Nationalist commander defected, and the Red Army closed, killing one-half million Nationalists and capturing one-third million more.

These battles provide us with the raw material to generalize first on classical tactical ploys and then on the evolution of modern tactics and their exploitation of and response to terrain characteristics.

REFERENCES

Chandler, D. 1987. *The Dictionary of Battles*. London: Ebury Press.

Creasy, E. S. 1852. *Fifteen Decisive Battles of the World: From Marathon to Waterloo*. Reprint. New York: Military Heritage Press, 1987.

O'Sullivan, P., and J. Miller. 1983. *The Geography of Warfare*. London: Croom Helm.

7

CLASSIC SPATIAL PLOYS

Strategy is the orchestration of battles in a war to achieve a political end; tactics is concerned with the management of a battle. Until the middle of the nineteenth century, battles were fought within the range of the vision of a commander. Closely arrayed groups of soldiers responded to orders, in the form of shouts, gestures, whistles, drums, trumpets or flags, with rehearsed movements or actions. The commander's function was to arrange and direct soldiers and to use weapons in any geographical setting so as to win. Tactics, then, consisted of using a variety of spatial ploys to kill enough of the enemy or to disrupt their collective integrity to the point of concession by flight or surrender.

DEFEND OR ATTACK

The fundamental tactical choice is whether to attack or defend. To attack is to take the spatial initiative and to accept the disadvantages of motion and commitment. To defend is to take the inherently stronger position. To pick the ground, take a stance, wait for the opponent to show his hand, and stand to take him on the move usually is more effective. Assuming an equality in weaponry—unless the defender has been geographically foolish in the place he chose or fails to keep watch on the attacker and gets surprised—an attacker will usually require a superiority in numbers and firepower in order to win. The advantage of the lie of the land usually accrues to the defender, which magnifies the power of his troops and weapons. The attacker accepts the potential cost of momentum. To commit one's force to certain direction is to take an irreversible chance. The great generals have often combined strategic aggressions with tactical defense. They employed superior mobility to force their enemy into a position where they had to decide whether to advance or retreat. Having jockeyed the enemy into a disadvantageous geographical position, it has often paid to wait for them to make the first move in battle. The great talent in fighting is knowing when to turn from the defense to the attack. The defender's advantage of superior position in the landscape will only pay off if he can turn on the attacker when his weakness shows. Most close contests are decided on the razor's edge between the attacker's advantage of surprise and the vulnerability of overcommitment to detection. Battles, like games, are won or lost in the transition between attack and defense. The decision to attack or defend logically should be a matter of assessing one's strengths and weaknesses in comparison with those of the opponent. One carries the fight to the opponent if one has enough force to overpower him and enough quickness to parry if one misses and he counters. If one is unsure of one's superiority in strength or speed, one trades space for time: let the enemy come on, give him the maximum opportunity to make a mistake, pick the ground carefully and contain the enemy's thrusts until a certain prospect for a counterattack shows.

Given that we are creatures of habit, the actual selection of tactics has usually involved doing what proved successful in the past. It has been a matter of the culture of the military caste and a temperament cultivated by tradition. The cavalry spirit and the convention that attack was the best form of defense, which arose among an officer corps whose chief experience had been against poorly armed people in a widespread colonial empire, produced the senseless slaughter of the Somme. The lessons that were plainly to be seen in the American Civil War and the Franco-Prussian War did not strike home. The deadliness of rifled firearms and artillery did not seem to have modified the tactical temperament. The same people who had used machine guns on Zulu and Sioux persisted in acting as if they were facing muskets. A defensive frame of mind also led to defeat if it pervaded the wrong time and place. The passive engineers' approach, which held sway in French military circles between the wars, the Maginot mentality, was out of step with the technological possibilities of planes and tanks.

It has been suggested that the geographical conditions of the polity to which they are attached has much to do with the military ethos. The empire that France acquired in the nineteenth century, with a few soldiers under a strongly centralized chain of command covering a vast area of lightly populated territory, produced officers skilled in constructing small defensive works and managing small units with a meticulous eye for detail. Their lack of experience of large-unit field maneuvering and their hesitation to take local initiative showed up in 1870 and seemed to survive till the 1930s. British insularity has been credited with creating a defensive tendency in Europe. Long supply lines and small armies bred a tendency to conservation. Wellington's eye for a position, his frugality with men's lives and his construction of defenses did not, unfortunately, survive as a style into the twentieth century. That same insularity supposedly made the sea the British element and fostered the aggressive Nelson tradition of seeking battle and the strategic doctrine of battle fleet supremacy. The vast spaces and deep winter of Russia are natural allies to its soldiers. Time and again they have won with an active defense, buying time with territory, spreading the enemy out to be ravaged by bitter cold. The defensive spirit still shows itself in the sword-and-shield formation, and even in the development of antiballistic missile defenses.

To have a known predisposition to attack or defend is, of course, foolish since it limits the options the enemy must prepare to face and it reduces the disturbing effect of uncertainty. It is obviously more efficacious to leave the choice until a calculation of the balance of advantage can be made, taking into account the circumstances of time and place of both parties to the conflict. A known inclination one way or the other allows one's opponent to concentrate on any shortcoming that has inevitably shown itself in the past.

Once the decision to attack or defend is made, the tactical choices are concerned with disposing the forces in the landscape so as to maximize their effectiveness and to minimize that of the opposition.

A choice to defend can be undertaken by making a stand or by retreating. A static defense involves trying to present an impenetrable barrier for the enemy to break against, while ensuring that a line for supply and retreat is kept open. The objective is to cause the maximum damage to the attacker while stopping him from getting through one's own lines or around one's flanks and, thus, behind one. Mobile defense takes the form of an orderly retreat, luring the attacker to extend his lines and reduce the concentration of his forces. By maximizing the length of his front and the territory behind it, his forces are drawn out as thinly as possible. This creates the opportunity to tear at the enemy's interior lines, ripping out inner flanks to gnaw upon them.

To decide to attack is to commit to movement. The aim is to penetrate the defense or to come at it from the side or back so as to destroy it more easily. By deceptive movement, the defense is thrown into confusion about what direction to face, and in this confusion weaknesses in the defensive line present themselves to the attacker. The offensive objective is to maneuver so as to produce a concentration of force sufficient to overwhelm the defensive deployment, taking it on a bit at a time.

Clearly, there is an enormous number of variations on the basic themes of attack and defense, with the same contestant taking on both rôles in the course of the battle. Add to this the variety of arms and the uniqueness of the landscape and it is evident that there is a large number of possible tactical permutations. However, the compact formations of soldiers employed before this century did make possible some geometrical generalizations on tactical ploys.

FORMATIONS

The tactical arrangement of soldiers can be described with three geometrical images: the line, the column and the square. Tactics can be reduced to the decision about which one of these to form in particular circumstances, when and how to change which way they face and move or when to redeploy from one formation to another.

To form a line is to present the greatest length of fighting frontage and the most difficult formation to outflank. Used in defense it is the deployment of containment, and in this can be reinforced with structural elements, trenches, walls, redoubts and obstacles. To advance in line on the offensive brings as much of any army's firepower or cutting edge to bear as possible. The capacity of a line to make outflanking difficult can be augmented by anchoring the line on a geographical barrier to movement. The sea, a lake, a river, bog, an escarpment, village or a town can be used as an end stop for a line drawn up across an avenue of approach, on the shore or a plain. The disadvantage of the line is its lack of depth. It spreads a force very thinly and, thus, is vulnerable to penetration. The line is also immobile. It is very difficult to maintain the integrity of a moving row of soldiers when they are strung out across the landscape. Off-road travel is not easy in any environmental setting. If one anchors a line on a geographical feature to prevent a flanking movement, the line becomes less mobile, either in attack or retreat, because of the need to keep contact with the end stop.

Turning a line to face one of its ends produces a column. This second formation gives up frontage to achieve fluidity. Since its flanks are linear, it is very susceptible to attacks from the side. The narrow front and the long, exposed flanks are the shape of an army on the move. Masters of mobility, like the Mongol Khans, Napoleon and Sherman, advanced in several columns, keeping the enemy guessing with speed and bringing the full weight of their forces to bear suddenly where the enemy was weakest and unprepared. According to Napoleon, "The strength of an army, like that of a moving body in mechanics, is expressed by the mass multiplied by velocity." The column is the formation with which to achieve velocity.

If one wheels a column in the same direction three times, so that it catches up with its own tail, and then face it outward one has formed a square. The square is the most static and coherent formation; it presents front in all directions and has no flank to be turned. It is the most defensive formation and in the limit becomes completely static where it is enhanced with fortifying structures. To go back to back is a most efficacious way of dealing with an enemy who has the advantages of both numbers and speed. The danger of an attack from behind is obviated. The price, however, is a lack of lines for supply or retreat. The square can be moved with careful coordination to maintain its dressing and present no gaps for penetration. Using the Mediterranean as one wall and foot soldiers as the others, Richard Lionheart marched from Acre to Jaffa in 1191, protecting his cavalry and baggage from Muslim archers in a combination square and column until Saladin offered battle and the crusaders' heavy horse could be brought to bear effectively against the Saracens. The static form of Wellington's infantry squares tore up Ney's cavalry in 1815. The built form of the Russian redoubts at Poltava did their job in raking the innards of the attacking Swedes.

AGGRESSIVE PLOYS

Once it has been decided to take the initiative in fighting and to attack, there are three basic prospects. One is to advance at the enemy's lines head on in a frontal assault. The second is to go around the end of the enemy's formation to smite them from the side or back in an outflanking attack. The third is to concentrate the attack on one part of the opposing line, trying to drive a wedge in to split the line and, effectively, to form two new flanks to attack. We can describe the main variants in classical aggressive tactics with reference to these elementary moves. Most battles consist of combinations of these tactics—deceptive feints, counterstrokes and the employment of the terrain.

If one has larger numbers or better weapons, discipline or aggressive spirit, the frontal assault may bring victory in the most direct fashion. If there is near equality in men and firepower, this can produce wasteful carnage.

To avoid this, the flank attack takes advantage of greater mobility to strike the enemy in the side or back. The most primitive form of flank attack is still the favorite tactic of guerrillas: ambush. After hiding where the landscape provides concealment close to an avenue of advance, an army can deliver a flanking attack to the enemy while they are on the move in a column. The objective in general is to roll the enemy's line up from one end or to sweep to their rear, cutting off the line of retreat and supply.

The usual combination has been a frontal attack to fix the enemy's attention with a hook to the side or rear. Outflanking developed naturally from a frontal attack with superior numbers, which enabled the attacker to curl around the end of the defender's line and roll it up. We have examined the first recorded contrivance of a flanking movement in the Theban refusal of the Spartan left at Leuctra.

Outflanking both wings of the enemy formation like a pincers is referred to as envelopment. Given similarity of numbers, it can be brought off by sucking the enemy in between the jaws of the pincers. His center can be refused with a feigned retreat so that he breaks ranks and gives pursuit, only to be struck from the side by one's waiting wings. The legions of Scipio

Africanus at Ilipa, von Moltke's Prussians, and Shaka's Zulu impis all employed a thrusting envelopment with their wings closing more rapidly to sweep around the ends of the enemy's line and draw them in on the concerted mass of the attacking center.

The wedge, employed to open a chink in the defensive line, concentrates the attack to achieve penetration. Effectively this is a variant on the flanking theme. Two new flanks are created so as to roll up the defensive line and threaten its rear. Alexander broke the Persians at Arbela with this tactic.

DEFENSIVE PLOYS

The line and the square represent the essence of defensive tactics. As we have seen, one can either make a defensive stand or defend on the move. Whatever combination of formations and movement is employed, the great advantage of the defensive is that it has a better chance to pick the ground on which to fight. The landscape can be turned to defensive advantage to provide screening and cover for one's own army and heavy going to break up the attacker's advance. Picking a place where the offensive is channelled onto ground that slows them up provides a killing field. The defense obviously has the best chance to use slopes, rivers, shores, bogs or settlements to augment their formation. Beyond this, the static defense has the opportunity to add to its topographical advantage by building fortifications and laying out fields of obstacles. Fortifications can either seek to hold back the enemy's advance along a long front in a line or to form squares to hold key locations in the landscape and break up the aggressor's momentum.

Mobile defense found a classical expression in Fabian tactics: cautious delay and avoidance of battle. A brilliant, fluid defense sustained the Byzantine Empire for 500 years. The mobile defense is the strategy of partisans and guerrillas, who overcome the superior numbers and firepower of a conqueror with speed, stealth, surprise and the cover and concealment offered by the terrain. Time is bought with ground, and the enemy's force is drawn out thinly so that it may be eroded by unpredictable attrition. The conqueror's forte, pitched battle, is avoided at all costs.

TACTICS TODAY

The next chapter discusses the development of tactics since the advent of rifling. The increased lethality of weapons eventually led to a departure from close order, and the line became a defensive zone with depth, taking advantage of cover and camouflage with strong points and tank traps in an expendable belt acting as a friction zone to slow and bleed an armored attack. This defensive cloud was matched by a racing flood, which replaced the offensive line. A torrent gushed through the weak points in the defenses, reaching in to disrupt the defender's interior lines. Modern aggressive tactics constantly seek to outflank the enemy and to get to his rear with mechanized columns churning in a variety of wedges, pincers and flanking movements, seeking any weakness in the extended, permeable fronts of mobile warfare. Add to this the dimension of aircraft which provide vertical flanking movements that strike from above and the deep penetration of perpendicular insertion of troops by parachute or helicopter.

The pervious formations and fluid thrusts of mechanized warfare with long-distance communications and air support have extended the battlefield and the duration of battle. Sprawling operations with sporadic firefights have replaced the set piece of the past. The movement and placement of formations to outfox the enemy are now strategic rather than tactical functions. One commander can no longer directly oversee a battlefield. The simple geometry of lines, columns and squares, designed to achieve coherent action in response to direct controlling signals, is neither necessary nor desirable today. Tactics, the art of disposing soldiers and weapons in the field within the perceptual scope of a commander, devolves to a very local task. It involves the detailed placing of small units to perform a given task so as to take advantage of geographical features and the actions and responses of the enemy in a limited section of the whole battlefield. The command of armies has become a strategic function. Tactics, then, is left largely with the task of training soldiers how to fight sensibly. They are instructed and drilled on how to use their weapons and vehicles to best advantage and how to avoid being killed by using the lie of the land to magnify the effectiveness of their firepower and to diminish the danger from the enemy's.

REFERENCES

Montross, L. 1960. *War through the Ages*. New York: Harper and Row.

O'Sullivan, P., and J. Miller. 1983. *The Geography of Warfare*. London: Croom Helm.

8

TERRAIN AND TACTICS

The last two chapters considered battles and generals' stratagems. Now we come down to the earth itself and the military exploitation of the features of the landscape. The basic fighting handbook of the U.S. Army, FM 100–5, says:

> Weather and terrain have more impact on battle than any other physical factor, including weapons, equipment, or supplies . . . Indeed most battles have been won by the side that used terrain to protect itself and to reinforce fires to destroy the enemy. To be effectual, commanders must understand the nature, uses and reinforcement of terrain.

Fighting consists of moving, hiding and shooting. Or as J.F.C. Fuller, one of the apostles of mobility, put it: "In the abstract, tactical warfare may be considered as a combination of three elements: mobility, protection and offensive power." The improvements in vehicles and weapons over the last century and a half have expanded the size of the battlefield and have placed a premium on detailed geographical knowledge of the battlefield. We introduced the elements of terrain analysis in Chapter 2. In the context of selected recent wars, we reviewed the notions of key terrain, avenues of approach, points of observation, fields of fire, and cover and concealment. In the light of the three elements of warfare—moving, hiding and shooting—these reduce to two basic considerations. Mobility requires an assessment of the going over a piece of landscape, the ease with which men and vehicles can move. Shooting and avoiding being shot call for an assessment of sight lines. Questions of fields of fire, observation and cover and concealment all come down to the measurement of how far one can see or range weapons from various places.

With these two characteristics of the battlefield in mind—the going and the sight lines—we can review the recent evolution of tactics in terms of technology and its geographical implications.

MODERN TACTICS

From 1700 to 1850 the European armies, which dominated the world with smoothbore guns and socket bayonets, deployed in lines of infantry three deep, backed by direct-fire artillery and flanked by cavalry. So equipped, 20,000 men were needed to hold a mile of battlefront. The development of breech-loading, rifled firearms in the middle of the century increased the range of engagement from 150 yards to 500 yards. This led to a spreading out into skirmishing lines on the attack, and to digging in on the defense. By 1870 12,000 men were sufficient for every mile of front. By 1917, with the introduction of the magazine-fed rifle, smokeless gunpowder, machine guns and indirect, quick-firing artillery support, 2,500 men could hold a mile of battlefront.

The emergence of the centralized nation-state, linked by railways and telegraphs, had meanwhile increased the speed and magnitude of mobilization and the control at the strategic level. The motor vehicle and telephone added to the tactical advantage of static, positional warfare. The battlefields of World War I were dominated by the defensive might of indirect artillery firepower and machine guns. This resulted in the stalemate of the trenches which reached from the North Sea to Switzerland.

The lessons from this meat-grinding waste, and the development of radio, light machine guns, mortars, grenades, tanks and aircraft, paved the way for a tactical emphasis on the mobile offense. There was a distinct departure from positional warfare and linear deployment and from the target it presented. Stress was laid on fluidity and decentralized command. The theoretical and practical precedents were established during the war. A Captain Laffargue had advocated the carrying of light cannon and machine guns forward with the attack to penetrate and get behind enemy lines. In 1918 the Germans gave a demonstration of "blitzkrieg without tanks." Searching out soft spots on the opposing lines as thrust points, attack teams infiltrated and disorganized the Allied rear. Meanwhile Allenby's campaign in Palestine demonstrated the superiority of highly mobile penetration and deep exploitation. Fuller's Plan 19, proposing a tank-led, thrusting attack, identified the enemy's command structure and center as the object of the exercise.

Between the wars, the apostles of mobility, Fuller and Liddell Hart, theorized on fluid, armored tactics. Fuller, who stressed armor, relegated the infantry to a mere holding rôle; Liddell Hart called for infantry in close support of armor to operate as an expanding torrent, flowing around pockets of resistance. The Russian and German armies went beyond theory to practice, elaborating doctrine, training and equipment for mobile warfare. While Tukhachevski left the Red Army a legacy of mobile operations, in his field manuals at least, Guderian developed the blitzkrieg. This combined infantry attack conducts battle over a broad band, seeking out points of weakness for exploitation with tank, artillery and dive bomber support. German training stressed dash, initiative and local tactical decision making. The counterpart of the offensive blitzkrieg was defense in depth, holding strong points in a wide zone, allowing the enemy to come on so that his flanks and rear could be attacked in an active defense.

It was against such a defense, and in the terrain and winter of the Soviet Union, that the blitzkrieg finally came to grief. These fair-weather tactics require good flying weather and good roads. In the forested and marshy northern section of the Russian Plain, where there was virtually no going off the road for tracked or wheeled vehicles, the armor of the *Wehrmacht* was roadbound. The Red Army moved through the boggy woodlands, forcing the Germans to fight an infantryman's war in the forests, where their tanks, artillery and air support were of no use. Soviet soldiers were at home on the land, tougher and better at close quarter fighting. Even before the bazookas arrived from the United States in 1942, the Soviets were using the Molotov cocktail, devised by Republicans in Spain, and the satchel charge to knock out tanks. The Soviets instinctively evolved a web style defense, filtering behind German lines to create pockets of resistance as bridgeheads for further extensions of their gnawing, aggressive defense. The blitzkrieg ground to a halt in the face of this and the winter of 1941. The same tactics, scaled down and modified for urban terrain, brought the Red Army victory in Stalingrad. A defensive web of strong points was established in buildings, allowing the Germans to penetrate the streets. Here, they were attacked on the flanks and rear, and their tanks and infantry were separated. In their active defense with small, independently operating storm groups, the Soviets engaged the Germans as closely as possible to avoid the *Luftwaffe*'s intervention. The swansong for German armor came at Kursk in July 1943. In its retreat, the *Wehrmacht* itself gave an amazing display of an active defense in depth. Although they usually tried the double envelopment, attacking simultaneously on the flanks and the rear, the Soviet armored offensive depended more on numbers than on finesse. In some instances, the German defense could be overwhelmed only with a superiority of 7:1, or more.

Elsewhere infantry came into its own where armor floundered in close terrain with ample cover, in the hills of Tunisia, along Italy's rugged spine and in the bocage of western France. In Normandy, with its small fields surrounded by hedge-topped, earthen banks and ditches, narrow lanes and scattered hamlets, tanks were at a decided disadvantage. Despite their command of the air, the Allied offensive required a 5:1 superiority in men and firepower to succeed. In the tropical forests of Malaya, Burma and the Pacific, victory was won by shrewdly employed infantry firepower. Although the Japanese had no greater experience of jungle fighting than the British, they adapted to the jungle in camouflage and tactics. The Japanese fought in the jungle, but the British stuck to the roads, which invited the employment of the roadblock tactic. Japanese units would circle through the jungle to get behind a British force on the road, attacking it from the rear and cutting off its line of retreat and supply. Under Slim in Burma, British infantry learned to use the natural setting to their advantage and turn the tables on the Japanese. In the Pacific, the U.S. Army under MacArthur gave a brilliant display of oriental indirectness. MacArthur moved through the Philippines and "hit 'em where they ain't," leaving pockets of resistance to "die on the vine." The U.S. Marines preferred the direct, full frontal assault on the grounds that the shortest distance between two points was a straight line. Marine amphibious, island-storming tactics employed the maximum of infantry-borne firepower.

The success of the People's Liberation Army in China in the 1940s was achieved with fluid, small-unit infiltrating tactics. These were displayed to great effect in the early 1950s in Korea against American forces. On the move, the Chinese avoided highways and flat terrain. Despite the images conveyed by the American media, movies and comics, the Chinese did not attack in human waves. Assault teams made a stealthy approach to enemy emplacements and attacked suddenly at night without artillery support, which would warn the opposition of their intentions. They used their tight-knit teamwork and fleetness of foot to great advantage.

Since 1956 a lot of attention has been paid to the Israelis who are relearning the lessons of World War II in their fights with Egyptians, Jordanians, Syrians and the PLO. Those who looked to the *Wehrmacht* as a model prevailed as a refined version of blitzkrieg emerged in the tank-led *ugdah* formation. Despite its success in 1967 in the Six Days War, its weaknesses were visible in the tendency for the tank spearhead to leave its infantry support eating its dust, offering the prospect of the enemy regrouping, which would sever the spearhead from its shaft. Improved tank-killing weapons in the hands of infantrymen have made armor more vulnerable, even in the sparsely covered setting of the desert. Ground-to-air missiles have reduced the advantage of air superiority. The production of hand-held weapons that can effectively down helicopters has added enormously to the potency of foot soldiers in combat with mechanized and armored foes. Stingers in the hands of the Mujahedin probably did more than anything else to persuade the Soviets to withdraw from Afghanistan.

These technical advances, which place more effective firepower at the command of infantry, enhance the current fighting supremacy of intelligently led, decentralized teams of foot soldiers. This supremacy was evident by 1945. According to S.L.A. Marshall (1947), the chief tactical lesson of World War II was "the overpowering effect of relatively small amounts of fire when delivered from the right ground at the right hour". Time and again, the tide of war was turned by a few men in the right place at the right time.

Timing is often a matter of coordination and communication from higher levels of command. But either a mobile, indirect attack or an active defense calls for decentralized tactical decision making over a broad band of territory with numerous terrain opportunities for success, or obstacles to victory. This places a premium on the ability to comprehend and exploit the advantages and to avoid the pitfalls of the lie of the land at the lowest level of leadership. This is quite unlike the circumstances of the positional war of 1914–18. Whatever terrain analysis was done came before the event with the planning of defenses at the highest level of command. With a continuous front and 2,500 men per mile of line, there was no room for tactical options and maneuver. The only objective was the stretch of the enemy line before you.

On the modern battlefield, where deadlier weapons mean that less than 1,000 men can hold a mile of front, the tactical balance has, however, tipped back to where it was in 1914–18, to the defense. Clausewitz would seem to have had it right when he called defense the intrinsically stronger form of combat. Aggressive use of armor and air support can be held at bay to a large extent by hand-held firepower. As we have seen in Normandy in 1944, it took five Allied soldiers to overcome one defending German. On the eastern front, the *Wehrmacht* repelled some attacks mounted with a 7:1 advantage. The conventional wisdom that today a 3:1 advantage is sufficient for an attack to prevail in open country may well be a serious underestimate of the defensive advantage. The Gulf War, where the coalition had total command of the air, may well be typical of American intervention in global affairs, but it hardly represents the circumstances of most deadly competition in the world. Where one side does not have a monopoly in the air, the advantage probably still lies with the defender.

TERRAIN ANALYSIS

The dispersal and velocity of mechanized warfare with radio communications and airborne firepower has greatly expanded the battlefield in time and space. Set piece battles fought between sunup and sundown are a thing of the past. The prospect now is of a sprawling zone of continuing, sporadic firefights which erupt day or night over a period of a week or more. The maneuvering of formations in broad sweeps to outflank the enemy has become a matter of strategy rather than tactics. The battle zone is too big for one person to keep an eye on what is going on and to direct the action from a lofty viewpoint. In the past, the formal geometry of simple, drilled movements in response to simple signals fulfilled the need for control and concerted effort. Such regularity is no longer necessary or desirable. Tactics, the art of arranging soldiers in the field of battle within the purview of a single commander, today dissolves to the details of deploying men and weapons to a given task so as to take advantage of opportunities presented by the lie of the land and the actions of the enemy within a very local scope. The command of armies becomes a matter of strategy. What is left of tactics today is found, for the most part, in the training of soldiers in commonsense ways of fighting, using their weapons and vehicles to best advantage on a piece of the earth's surface and avoiding the enemy's deadly fire.

Given the task of stopping an enemy advance over a piece of ground, or of taking it away from him, today's soldiers seek to emulate water flowing over a surface of military opportunity and obstacles. Units surge along open channels, sweep around resistant obstacles and eddy into sheltered pools. Their object is to seek the line of least resistance on the offensive, or to soak the landscape so as to bog down the enemy's advance on the defensive. For either advance or defense, a thorough appreciation of the terrain is crucial to success. As a first step in this, it is necessary to identify the places one wants or wishes to deny to the enemy. The search is for natural or man-made features that give one an edge in combat. The most desirable places are those that give a sheltered view of the theater for observation and fire. Ground that is exposed to enemy sight and firepower is to be shunned. Having determined where to go, the next step is to find how to get there with the minimum effort or loss of life. The defensive equivalent is to deduce how the opposition might approach their presumed targets. In either case, it is necessary to assess how easy the surface will be to cross. The routes available and the natural or artificial obstacles that inhibit movement of men and machines must be surveyed. This involves not only the quality of the roads and how the quality will change with the weather; lines of movement to and from the targets and the speed with which they can be travelled must also be established. The analysis of terrain, dividing the landscape into places to be possessed and places to be avoided and specifying the access offered by routes and the lie of the land, provides the basic information for maneuvering and deploying men and weapons. Beyond this, the detailed choice of particular paths and placements is a matter of fitting arms and vehicles to the landscape and of second guessing the opposition.

The fundamental tactical level is the battalion command. This operates over an area beyond the scope of individual perception. At this level, units are assigned to the landscape by remote command in what is in essence an exercise of map reading skill. The next level is the company, where the commander fights the battle on the basis of what he can see. The elementary level is that of the platoon or squad. Here the emphasis is on the actions of the individual soldier.

Platoon-level terrain appreciation mostly involves getting soldiers to use cover. They are trained to seek frontal cover and to avoid landmarks, walking on the skyline and kicking up dust. They are encouraged to move quickly in the open, having marked cover they can run for if attacked. They are admonished not to mask their covering fire. Particular attention is paid to the limits imposed on the performance of weapons by terrain. For example, the missile fired from a tube-launched, optically tracked, wire-guided weapon (TOW) takes five seconds to travel 1,000 yards, and it takes some time for the operator to latch onto a target and fire. Thus, for an engagement at the range of a mile, a tank must remain exposed for 100 yards or so to be vulnerable. If it can be driven behind a hill, building or clump of trees quickly, it is safe. The weapon is best brought into play where the terrain allows constant visual contact. Woods, hills, buildings or dust can nullify its value. The TOW is best used in flanking fire from a hidden position at a range of from 2,000 to 3,000 yards. This standoff is the distance at which the TOW can be used to engage a tank and yet remain out of its range. Such detail provides clear tactical guidelines for deployment in the landscape.

The company commander's main concern is with enemy avenues of approach and the location of supporting troops. In assigning men and weapons to battle position, cover and commanding locations are sought for long-range antitank fire and shorter range flank and crossing fire, disposed so that obstacles impede the enemy's movement through fields of fire. It is possible to be very precise about what will slow up a tank. Tanks have trouble with slopes over 30 degrees, sharp rises over 5 feet, gullies or ditches over 15 feet wide, rivers or canals over 150 yards wide, bogs deeper than 3 feet, forest with trees 8 inches in diameter, trees 4 inches thick on a 10 degree slope, tree stumps 18 inches high, snow 3 feet deep, or a landscape of closely spaced buildings.

Because the battalion commander must direct action beyond the eyeball's range, he must rely heavily on maps. This is the level at which the enemy's and one's own key objectives in the landscape must be identified. Paths to these and obstacles before them, especially for armor, need delineation. On the defense, the alternative paths to the enemy's likely targets are ranked according to the likelihood of their use, and forces are allocated among these according to the potential danger. For each avenue of approach, observation points, fields of fire and cover and concealment are evaluated to deploy the firepower allocated to the best advantage. On the attack, the same analysis would apply, inverted to uncover any weakness in the likely deployment, and to seek to gain the advantage of surprise.

MAPS

The increasing range and deadliness of weapons through this century has induced not only an increase in the dispersion of troops, but also a decrease in the appropriate map scale for tactical decision. In 1918 company and battalion commanders used 1:25,000 scale maps. In the 1940s and 1950s the standard small unit scale was 1:50,000. In Vietnam, airmobile company commanders were forced to use 1:250,000 scale maps because their helicopters flew across the expanse covered by a 1:50,000 sheet before they could locate themselves on it. The dispersion forced by the use or threat of tactical nuclear weapons or gas and the speed of modern mechanized warfare also encourage the use of small-scale maps by company-sized units.

The kind of information needed from a map reflects the command scope of the user in space and time. Those in command of corps and armies, like the generals of the nineteenth century, are concerned mostly with roads. They do not direct fighting in a tactical sense; instead, they allocate and move units into or out of battle. Their decisions have a lead time of hours or days before they influence the course of a battle, because of the time it takes to move major units, even if they are motorized.

The command of divisions or smaller units gets down to tactics and requires more detailed, more timely information. At this level it is necessary to determine fields of fire for line-of-sight weapons, concealment and cover from enemy observation and fire, going off the road for tracked vehicles, landmarks for helicopter routes and good landing zones for helicopters. At the tactical level, decisions may have an impact within minutes; at the company level, this may come down to a matter of seconds. In this setting, the military aspects of the terrain may change as quickly as the disposition of enemy and friendly forces. An artillery barrage can convert a cluster of houses at a crossroads to a tank trap of rubble, or a readily penetrable forest into a tangled barricade. What was a meadow can be turned into a quagmire by the passage of a tank company.

The increasing importance of terrain features and, thus, of maps to mechanized military operations has made the impact of erroneous or out-of-date maps potentially disastrous. In the U.S. invasion of Grenada in 1983, the Marines used out-of-date British maps; the Army actually used tourist maps. One result was that a marine airstrike hit an army command post. Cultural features, such as roads and the extent of the built-up area, may change dramatically between the time of survey and the time of use. On a battlefield, bridges disappear, dams are breached and buildings are reduced to rubble at a speed faster than changes can be posted on the best of maps. In Vietnam U.S. forces began to use recent air photos rather than maps for terrain analysis because of the rapidly changing face of the landscape in battle zones. A considerable effort is being made, at least by the U.S. military, to develop digitized, video-displayed, automated maps, which can be constantly updated and which have some built-in automatic terrain analysis capabilities. This is the chief research function of the geography department at West Point, for example. How well these would function in the heat, dust and grime of battle remains to be seen.

VARIETIES OF TERRAIN

Chapter 3 addressed the variation of terrain conditions with climate. Up until the winter of 1989, the big battalions of NATO and the Warsaw Pact were geared up to refight the closing stages of World War II where it ended on the temperate plains and hills of *Mitteleuropa*. Mountains, deserts, snowfields, forests and cities were regarded as unusual combat conditions requiring specialized troops, training and tactics. As the likelihood of fighting in the cultivated midlatitudes diminishes, the prospect of conducting battle in what have been regarded as extreme circumstances increases.

The steep slopes of mountain country limit even tracked vehicles to a few twisting roads. For foot soldiers, the going off the road is tough and exposed. The confinement of passage through mountain ridges to a few passes has allowed a few men to hold or delay large forces. Superior numbers cannot be brought to bear in a narrow defile, be it the Salang Pass in the 1980s or Thermopylae 2,460 years before. Even helicopters, which offer the prospect of vertical envelopment, did the Soviets little good in the Hindu Kush. The Stinger, which tipped the balance once more to the indigenous defense in a mountain setting, proved to be a great leveller. Mountains have always favored lightly armed foot soldiers over armored heavy troops, be they Austrian knights in Switzerland, Xerxes Persians, or Russian tanks in Afghanistan.

Deserts have the reputation of allowing ship-like maneuvering by armor on an unrestricted sea of sand. The reality is otherwise. Desert terrain rarely permits easy off-road movements by tracked or wheeled vehicles, and it is easy to get lost in. Loose sand and the sharp slope changes of arid erosion make the going difficult. Angular hillsides and steep wadi banks create obstacles and narrowly channel traffic. It was relatively easy for both the British and the Axis armies to dig in defenses against tank-led advances in North Africa in the 1940s. The appearance of extreme fluidity of operations in the Western Desert was more a matter of small armies tearing around an extensive theater, than that of an ocean-like battlefield. In the battles between Egypt and Israel in the Sinai Peninsula in 1956, 1967 and 1973, despite the vast acreage of the region considering the number of combatants involved, the fighting was over road junctions or passes. Apart from clogging engines and weapons, the clouds of dust raised by vehicles give the opposition ample warning of advancing armor. The lack of vegetation reduces concealment, and fields of fire are not cluttered by trees, bushes, crops or fences. In the Yom Kippur War in 1973, Arab and Israeli tanks engaged and destroyed one another at ranges of one and a half miles. This is not the spacing of war in the greenery of the temperate zone. Intense heat melts radio wires, shorts electronic circuits and makes metal untouchable. Fine sand grates through the sprockets of tank treads and wears helicopter blades at an alarming rate. The seasonal timing of operations becomes crucial and the cost of overcoming the environment enormous.

War in polar conditions does not seem to be a serious possibility at the time of this writing, but in the realm of international rivalry it seems that very few options for conflict can be dismissed entirely. Obviously, fighting in mountains and winter snowfields is a distinct prospect. Winter snows bring the Afghani conflict to a seasonal standstill but the spring brings a renewal of fighting. War in extreme cold presents very severe logistical problems, as well as unique tactical problems. In the summer of a tundra climate, the thawed ground above the permafrost is a boggy mess, prohibiting off-road movement. Frozen ground, rivers and lakes in winter make for much better going and maneuverability. Deep snow may, however, keep vehicles on the road. Infantry trained and equipped to ski or use snowmobiles have a decided edge.

Forests and cities are customarily lumped together by the military under the heading of close terrain. Open terrain consists of farmland, grassland or desert. Open terrain is evidently the preferred scene of war for the military establishments; close terrain of forest and built-up areas is something to be avoided if at all possible. The division into close and open terrain reflects the characteristic length of engagement in various landscapes. Close terrain implies fighting at close quarters; close quarters generate high casualty rates, limited movement and loss of central control. War breaks down into uncoordinated fighting. Lines of sight and runner communication, as well as radio transmission, are disrupted in forests and towns. Heavy firepower at short ranges usually results in higher casualty rates than when similar forces engage at longer ranges in open terrain.

Obviously the category close terrain blurs the very real distinctions between varieties of forest and townscape. The various forest types, including rain forest, jungle, temperate woodland and taiga, call for different equipment and different tactical responses. Neither is the urban scene a homogeneous habitat, warranting a singular adaptation. We shall return to this theme later.

KEY TERRAIN

The most desirable ground in the battlefield—one that provides its holder with the upper hand—varies with the nature of the habitat; the equipment, weapons and number of combatants; and their ultimate objective. The same feature may be highly attractive to one force, but something to be avoided at all costs by their opponents. For example, a curious inversion of the military landscape occurred in the highlands of central Vietnam in the 1960s and 1970s because of the use of helicopters by one side in the conflict. The Vietcong avoided the grassy hilltops of the region because of American domination of the skies. Instead they held the wooded hillsides, where they were hidden from aircraft observation. The air cavalry of the United States and the Army of the Republic of Vietnam (ARVN) needed the open hilltops as helicopter landing zones. Instead of the defenders holding the high ground and forcing the attackers to come at them from below, battles usually started with helicopters landing on a hilltop and troops attacking downhill trying to take Vietcong camps. The screening offered by their brush made hillsides key terrain for the guerrillas; the open hilltop landing zones were key to the air cavalry.

In general, however, the high ground still presents a defensive advantage, providing a platform for observation and fire to dominate the battle zone. With tank guns and antitank weapons that can hit and destroy targets over a mile away, commanding heights to see and fire from are once more at a premium. Modern weapons are so accurate, long ranged and destructive that anything that can be seen on the battlefield can be engaged, and anything that can be engaged can be destroyed.

With the increased accuracy and deadliness of weapons, cover and concealment becomes a primary consideration. Landscape features that limit a tank's line of sight, range of engagement and mobility are highly desirable for infantry. What is attractive to foot soldiers in terms of providing shelter may also be desirable to tank crews trying to avoid antitank weapons. Even helicopters now seek terrain cover by flying "nape" of the earth, just above the ground, below ridge lines and lower than the canopy of forests they fly past.

In the harshness of desert aridity or winter cold, the key to success is survival. Places of refuge in these hostile environments become the greatest geographical prizes. To ancient and modern armies, oases have been key terrain. In the cold of high mountains or on wintery plains, any shelter against the cold of night might mean the difference between life and death, and possession of the meagerest of huts has been hotly contested.

Beyond the tactical level, for corps and army commanders, the key geographical elements are routes. Their primary concern is moving large units around their theater of operations. Junctions and bridges or passes are their key terrain. Since major route intersections, terminals and crossing points are often set in an urbanized landscape, clearly built-up terrain cannot be avoided altogether in the course of war. Inasmuch as towns command transport, communications, economic and political transactions they are the dominant features in the landscape, the chief prizes of war and, thus, the ultimate key terrain. If they are surrendered without contest, one might say that good sense has prevailed. If they are not, the decisive action may be bloody and costly to both sides.

REFERENCES

English, J. A. 1981. *On Infantry*. New York: Praeger.

Fuller, J.F.C. 1925. *The Foundation of the Science of War*. London: Hutchinson, p. 148.

House, J. M. 1984. *Towards Combined Arms Warfare: A Survey of Twentieth Century Tactics, Doctrine and Organization*. Ft. Leavenworth, Kans.: U.S.

Army Command and General Staff College, Combat Studies Institute. Research Survey No. 2.

Marshall, S.L.A. 1947. *Men Against Fire*. New York: Morrow, p. 68.

Liddell Hart, B. 1967. *Strategy*. New York: Praeger.

O'Sullivan, P., and J. Miller. 1983. *The Geography of Warfare*. London: Croom Helm.

U.S. Department of Army. 1982. *Operations*. Washington, D.C.: U.S. Government Printing Office. Field Manual 100–5, Section 3.1.

9

GUERRILLAS AND COUNTERINSURGENCY

With the standoff between the United States and the Soviet Union between 1945 and 1989, many of the 22 million war deaths and much of the destruction of human conflict during this half century have been wreaked in little wars fought in the shadows of the shatter belts. Local, colonial or civil wars have been waged in which guerrilla tactics have been used by one or both sides. Since World War II regular forces have been used primarily in counterinsurgency operations.

Guerrilla fighters are motivated not by professional pride, a sense of duty or drill-induced reflexes and a fear of punishment, but by religion, nationality, bravado and camaraderie. They have achieved considerable success as an auxiliary or prelude to the fielding of conventional formations, or as propaganda by action to exploit political weakness, to create a power vacuum and an elite cadre to fill it. There is no clear-cut division between regulars, guerrillas and terrorists in the political use of violence. Conventional armies employ units that use guerrilla tactics in a disruptive capacity as well as terror as a weapon. Revolutionaries adapt regular military formations as soon as numbers and arms permit. Yet there is a widely recognized variety of fighting in which irregulars combine military and psychological objectives in their operations and apply a highly developed geographical sense to exploit the advantages of terrain to discomfort the established order, be it a government or a conquering army. By contrast with conventional war, there is a sharp distinction between the possibilities and problems facing each side in such a conflict. The geographical opportunities favoring guerrilla action are quite different from those that make counterinsurgency operations easier.

GUERRILLA TACTICS

The essence of guerrilla tactics is to trade space for time. The enemy is allowed to dominate a lot of ground, but his morale and force are slowly eroded by a thousand small cuts. He is drawn to extend his supply and communication lines and spread his firepower thinly, so that his internal connections as well as his flanks may be gnawed and his resolve eroded by constant nipping. Hit-and-run raids, diversions, sabotage, terrorism and ambush are the chief engagements. Although feinting and running to avoid pitched battles are primary ploys and although strategically guerrilla war is usually defensive, to achieve success, it must be tactically on the offensive. Employing what Liddell Hart calls "indirect methods," guerrilla actions inspired his advocacy of waging a war of movement and surprise. In effect Hart turned the offensive defense of traditional guerrilla war on its head to produce his prescription for regular armies: the defensive-offense of deep penetration, in which the perfection of strategy would be to win without any fighting. The tactics of Hart's dispersed strategic advance and guerrilla warfare are best summarized in the slogan that Mao adopted from Sun Tzu: "Uproar in the east, strike in the west" (Griffith 1963).

The ultimate objective of both sides in guerrilla war is control of the people. Thus, in order to succeed, any force countering guerrilla action must not only defeat the insurgents militarily, but also politically separate the population's sympathy from them and ensure the existence of an acceptable social order and government. A campaign to "search and destroy" must be combined with a reasonable prospect of "winning hearts and minds" and sustaining a viable political authority. The best tactics to employ in countering insurgency depend on how embedded in the terrain and the social fabric the guerrillas are. To dislodge the incursion of a small band in an alien setting, such as Guevara's fatal adventure in the Bolivian chaco, a mobile strike force may be sufficient. In the case of an indigenous movement with some popular following, such as the FMLN in El Salvador, flying columns should be combined with an effort to control territory. The conflict within the Salvadoran military, between advocates of aggressive patrolling and advocates of massive sweeping operations, suggests that some combination of territorial control and mobile striking power is required.

If we consider the fighting component of counterinsurgent campaigns first, the main need is to get sufficient firepower to overwhelm guerrilla bands wherever they can be found, at rest or in action. This implies a small fast-moving unit—the flying column. The military component of the French "oilspot" strategy for countering colonial resistance used quick, ever-moving units, seeking friendly ties and intelligence from the population and employing "whirlwind" tactics to fight. The other elements of the oilspot approach were holding territory and pacifying the population. The military action was only a necessary prerequisite; in itself, it was not sufficient to attain the desired end state. Once cleared of insurgents, an area was taken over by the civil authorities who won over the locals with protection and the prospect of economic and social progress. The oil dropped on these spots spread French influence slowly and methodically over the land, according to the theory.

The British in South Africa, harried by the superior tactics of the Boers, used a more brutal method of controlling the civilian population, which was providing support to the commandos. The Boers' scattered farms were razed, and the noncombatant population was gathered in concentration camps. In Ireland, the British devised methods to screen the population for gunmen, which evolved in Palestine into "cordon and sweep" operations. This use of troops did little to endear them to the civilian population and, if anything, helped guerrilla recruiting.

In the light of these tactical and strategic considerations, can we generalize on the settings in which guerrilla warfare may be successful, or otherwise? The quantity that is bandied about in discussions of such warfare is the ratio of regulars to guerrillas. The conventional wisdom of the military and media is that it is necessary for regular forces to outnumber insurgents by at least 10:1 in order to stand a chance of defeating them. In 1965 Maxwell Taylor upped the ante, claiming that a superiority of 25:1 would be necessary to defeat the Vietcong, in recognition of the low proportion of teeth to tail of the U.S. forces.

To investigate the geographical dimension of this ratio, we can turn only to the historical record. This is rather spotty in its coverage and is subject to wide and variable margins of error. Errors arise not only from a lack of information but also from political deceit. Mike Wallace's accusation that General Westmoreland suppressed CIA data on the Vietcong to influence President Johnson not to withdraw in 1968 was never really refuted.

What data there are do confirm the economy of guerrilla fighting. It has, indeed, usually required a superiority of greater than 10:1 for counterinsurgents to win, but the actual ratio varies greatly from place to place. The obvious geographical variables that might come into play in determining the effectiveness of the two sides in guerrilla wars are (1) the availability of targets and support and (2) the texture of the landscape, offering cover and concealment.

POPULATION DENSITY

Taking the density of support and targets first, it would seem that the number of guerrillas needed to dominate combat in a given area would be larger the greater the number of targets and the greater the complexity of the human landscape to be controlled. The population provides at the same time potential material and moral support for guerrilla campaigns and also the density of targets for a war which must register as a constant irritant

Table 9.1
Population Densities and Regular-to-Guerrilla Ratios

WAR	Population Density (people per square mile)	Regulars per guerrilla
Guerrilla Wins:		
Algeria 1954-61	10	19
Cuba 1957-59	132	15
Indochina I 1945-54	180	2
Palestine 1945-48	167	2
Regular Wins:		
Kenya 1952-56	24	5
Uruguay 1969-72	39	11
Greece 1944-45	154	11
Cyprus 1955-59	142	21
Malaya 1948-52	132	29

to the incumbent authority. The number of regular troops required to counter guerrillas will also be greater at higher population and target densities. Since counterinsurgency requires comprehensive control rather than the sporadic exploitation of advantage of guerrilla tactics, however, the number of regulars needed at higher densities will be more than proportionally greater.

As a first rough cut at quantification, we can tabulate the ratio of regulars to guerrillas and the national population densities for several little wars fought since 1945 (see Table 9.1). There are too few observations to subject to statistical testing of hypotheses, but they do suggest that the generally greater effectiveness of guerrilla manpower is more potent at higher densities, where targets and support are potentially greater. In order to suppress guerrillas in high-density conditions, regulars must go beyond the 10:1 mark. The shift of revolutionary emphasis to towns makes sense in these terms, as well as those of economic and social structure.

TERRAIN

Military lore has it that the offensive defense of guerrillas has an advantage over regular formations in wooded and rugged terrain. This has intuitive appeal. Guerrilla tactics include hit-and-run raids and decoying and ambushing the enemy. These tactics place a premium on the availability of cover and concealment and accidented landscape to fade into, one which slows down and fragments the pursuit of conventional forces. From this we could deduce that the number of guerrillas needed to dominate a given amount of territory will decrease as the amount of cover available and the ruggedness of the land increase; in other words, guerrilla effectiveness increases with the intensity of vegetative cover and with slope of the land. By contrast, the flatter and barer the terrain and the less textured the landscape, the fewer are the regulars needed to comb it and establish command. Thus, counterinsurgency effectiveness could be expected to decrease with greater slopes and higher rainfall, which gives greater vegetative cover.

If we accept this premise of the argument, we would expect that regular, counterinsurgency forces would have to be deployed at a ratio greater than the oft-cited 10:1 in order to defeat guerrillas in broken country with lots of vegetation. A reasonable deduction from this is that, *ceteris paribus*, guerrillas have a greater chance of overcoming the incumbent regime in a broken, wooded landscape and that counterinsurgency campaigns are likely to defeat guerrillas in flat, bare country.

Clearly, other things are seldom equal and the human aspects of the landscape, as well as the political milieu and the geopolitical circumstances, come into play in determining the outcome of any violent conflict. Differences in the density of human occupation of the landscape, the spacing of settlements and the intensity of the road network may all play a part favoring one side or another tactically. Above and beyond these local geographical factors, the longevity or outcome of an insurgent movement may depend to some extent on the degree of popular sympathy for the rebel cause or the government. At the geopolitical scale, the support of a powerful nation and the availability of cross-border sanctuaries might influence the outcome in favor of the guerrillas.

Scott Rezendes (Rezendes and O'Sullivan 1988) tested the veracity of these deductions from the lore using inferential statistical methods. These were applied to historical evidence of the outcomes, numbers involved and geographical conditions surrounding the little wars in the shadows. Guerrilla mythology was expressed as quantified conjectures and subjected to trial by attempted statistical refutation.

His object was to test the hypothesis that the vegetative cover and ruggedness of the theater of war had a significant effect in favor of guerrillas in conflict with regular forces, taking into account the influence of human geography, the political atmosphere and the geopolitical circumstances on such strife.

Data

The only recourse available is to historical data. Quite obviously, to hold the effect of technology on the balance of advantage in different environ-

Table 9.2
Guerrilla Campaigns Since 1945

Conflict & Duration	Guerrilla win = 1 loss = 0	Regulars per guerrilla	Slope (%)	Rain (inches)	Roads per mile^2	Settlement per mile^2	Cross border sanctuary	Popular support	Foreign support
Aden 63-67	1	4	9	2	.17	.005	1	1	1
Algeria 54-62	1	20	6	30	.17	.005	1	1	0
Angola 61-75	1	5	2	39	.08	.001	1	1	1
Bolivia 66-67	0	11	9	38	.08	.003	0	0	1
Cuba 57-59	1	133	7	50	.31	.012	0	1	0
Cyprus 55-59	1	100	6	12	.33	.023	0	1	0
Greece 45-49	0	10	12	27	.26	.015	1	0	1
Guatemala 64-67	0	53	17	54	.19	.011	1	0	0
Guinea 59-75	1	4	1	78	.12	.008	1	1	1
Indonesia 47-49	1	2	6	70	.22	.009	0	1	0
Iraq 45-75	0	8	13	14	.15	.004	1	1	1
Kenya 52-56	0	4	6	35	.16	.006	1	0	0
Malaya 48-56	0	30	8	94	.10	.005	0	0	0
Mozambique 61-75	1	8	2	47	.11	.002	1	0	1
Nicaragua 77-79	1	5	7	57	.25	.013	1	1	1
Oman I 57-59	0	7	15	5	.11	.006	0	0	0
Oman II 63-76	0	8	8	5	.05	.002	1	0	1
Peru 62-65	0	40	14	25	.13	.007	0	0	0
Philippines I 46-54	0	3	9	79	.17	.014	0	1	0
Philippines II 72-79	0	2	8	79	.11	.024	0	0	1
Rhodesia 63-80	1	2	4	41	.13	.003	1	0	1
Sudan 55-72	1	3	1	35	.08	.003	1	1	1
Vietnam I 46-54	1	3	9	86	.17	.013	1	1	1
Vietnam II 60-75	1	7	6	64	.14	.006	1	0	1
χ^2 critical value = 3.84	—	0.06	13.58	0.74	0.74	0.01	5.63	8.20	1.38
Point biserial correlation, critical t = 1.70	—	0.47	4.23	0.47	1.32	0.30	—	—	—
Discriminant Analysis F to enter = 1.0	—	0.25	18.92	0.25	1.83	—	2.58	11.47	1.35
Regression on duration significant F = *	—	0.11	−0.71	0.42	−0.42	1.80*	—	1.1	2.47*

mental settings constant, one must focus on a limited era of similar technology. Given the availability of information, it was appropriate to consider only those wars fought since 1945. Setting aside the conflicts with indeterminate outcomes or no reliable information, this left twenty-four observations (see Table 9.2).

For the dependent variable of the analysis, it was necessary to determine who won and who lost. There are disparities between military campaigns and the final political outcomes of rebellions. Although the warriors of one side or another might claim victory in the field, the ultimate measure is political success. Military claims are subject to interpretation. There is usually no doubt about who ends up on the throne. For this exercise, winning is defined politically. There is, however, an aspect of the viability of a guerrilla rising that might be captured with an interval measurement. Longevity, or at least survival, might be taken to indicate a degree of success. Values for the duration of campaigns can be taken from the historical record.

To measure the physical and human characteristics of the terrain, the arenas of conflict had to be delimited. News sources and the literature were used to draw boundaries enclosing the principal battlegrounds.

The independent variables for this analysis were vegetative cover and ruggedness of the terrain. Vegetation is a response to climate, and rainfall is the most important determinant of cover in the populated latitudes where most political conflicts have occurred. The mean annual rainfall for each of the theaters of war in question was taken from meteorological records. As a proxy for ruggedness, the mean slope was measured within the area of conflict. A one-inch grid was laid over a 1:1,000,000 map covering the relevant region. For each grid intersection, slope was determined over a standard radius of four miles in the direction of maximum slope. The average slope was calculated from these values for the entire area.

To control for factors other than the physical environment in the analysis, it is necessary to have measurements of pertinent control variables describing the setting and events.

The artificial aspects of the terrain were deemed potentially important, and to capture these the density of settlements and the density of routes were measured within the conflict areas. All settlements within the boundaries were counted and divided by the areas in question. All discernible avenues of approach—roads, railroads and tracks—were measured and expressed as a density.

The local political situation was summarized with a binary variable indicating whether the literature suggested strong popular support for the rebels or not. Another, possibly ambiguous, expression of popularity is the ratio of regulars to guerrillas at the time of maximum involvement. This too was taken from historical accounts.

The geopolitical circumstances of the conflict were summarized with two more binary variables gleaned from historical description. Whether or not the guerrillas enjoyed a cross-border sanctuary could be fairly clearly established. Whether or not they had the support of a wealthy and powerful country, either a regional or global power, seemed relevant, if a little more difficult to determine with confidence.

Tests

In Figure 9.1 the distribution of guerrilla wins and losses is plotted between the two axes of slope and precipitation. This visual display gives a first indication of any regularity in the associations involved. The only discernible pattern in the array appears to run counter to intuition and conventional wisdom. Most of the guerrilla wins were registered in regions with less than average slopes; losses occurred in areas of steep slopes. There is no obvious relationship between wins and losses and rainfall and, by implication, vegetative cover.

Given the dichotomous nature of the dependent variable in this exercise,

Figure 9.1
Guerrilla Success, Rain and Slope

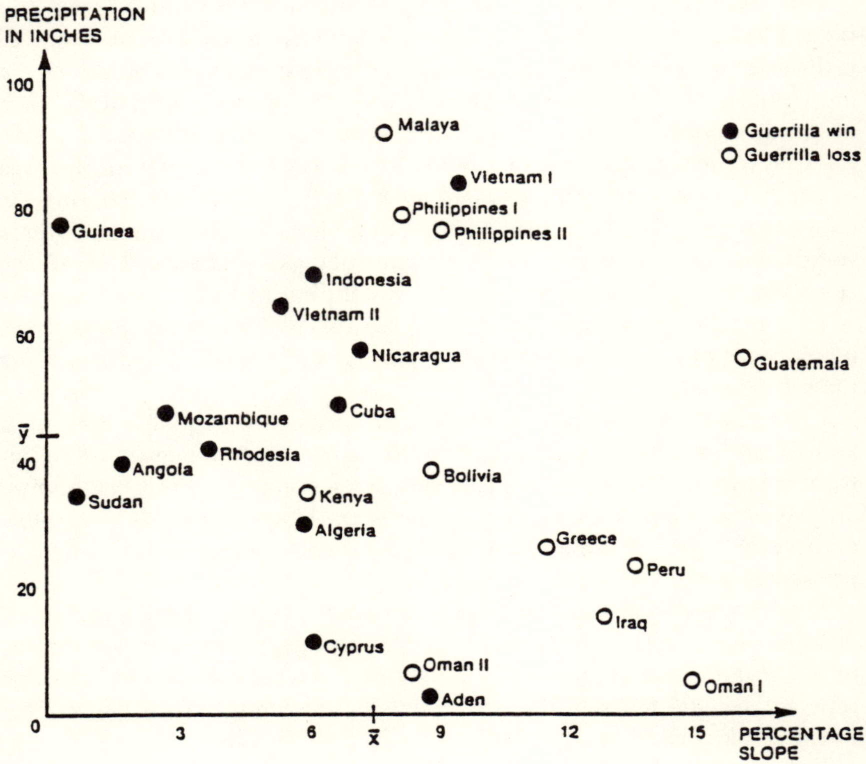

the first, most obvious step was to apply a χ^2 test to the frequencies of wins and losses arrayed against a series of dichotomized explanatory and control variables. For this procedure, those independent variables that were not already in binary form were put into this form. Observations were given a value of one or zero according to whether they were above or below average. For each independent variable, observations were cast into a four-way table according to whether they were wins or losses and high or low. The summary frequencies thus established were tested for the significance of their association. A step beyond this, to respect the interval scale on which some of the explanatory variables were measured, point-biserial correlations were calculated between these and the binary dependent variable. Perhaps the most appropriate method of analysis, however, for this mixture of dichotomous and continuous variables is discriminant analysis. This seeks the multivariate function that best discriminates between winners and losers. The significance of hypothesized determinants and control variables in separating wins from losses can be established. As a final exercise, to ensure that a different perspective does not uncover a radically different view, an interval measure of guerrilla viability, the duration of the conflicts, was employed. This was regressed as the dependent variable against the physical terrain variables under test and the control variables, both continuous and dichotomous.

Results

The χ^2 tests revealed that, of all the hypothesized determinants of guerrilla success under scrutiny, only slope had a statistically significant association with winning; however, it was in opposition to the presumed relationship. High slopes turn out to be related to guerrilla losses and low slopes to guerrilla wins. This was the strongest association with any potential explanatory variable. Rainfall, as a proxy for vegetation, proved to be insignificantly related to winning or losing. The control variables that proved to be significantly associated with winning were popular support and availability of cross-border sanctuary. The associations were not so strong as that for slope, but the directions of the relationships were as expected. The presence of popular support and sanctuary are significantly related to guerrilla wins.

Calculating a series of point-biserial correlation coefficients between wins and losses and the continuous physical and human landscape variables revealed only slope to be significantly related. Here again, the relationship was in the opposite direction to that hypothesized. Guerrilla wins occurred where slopes were lower.

The discriminant analysis was carried out with all of the potentially discretionary variables, the physical terrain measures as well as the geographical and political control measures in contention. The factors that best distinguish winners are, first, low slopes, then high route density, the availability of cross-border sanctuary and the enjoyment of popular support. No other variables were significant.

Using the duration of the conflict as a dependent variable, the same physical measures were tested for a significant, linear relationship, along with the geographical and political control variables in a stepwise multiple regression analysis. In this case neither of the proposed physical determinants of guerrilla viability proved significant. The only variables significantly related to long guerrilla campaigns were the availability of support from a foreign power and low village density. Between them these only explained 36 percent of the variance in the length of wars.

The results of the measurements and tests reported here confound a hypothesis derived from conventional wisdom. Convention has it that guerrillas do best in rugged and forested terrain. Turning this into a quantified proposition and testing its statistical validity with historical data not only denied but overturned the hypothesis, as far as the lie of the land is concerned. Since World War II, guerrillas have won most frequently in regions of less than average slopes. Vegetative cover, as proxied by rainfall, has no discernible bearing on the outcome of little wars. The slope factor proved to be consistently significant through a battery of tests, except when duration of the wars was used to measure guerrilla viability in a linear regression model. Among the methods used, discriminate analysis is the most appropriate for the task in hand. The significance of the negative relationship between slope and winning is clear in this. The control variables do, indeed, turn out to be of secondary importance according to this test. Those that proved to be of some significance include both the elements of the human landscape—road density—and elements of the local geopolitical setting—popular support and the availability of sanctuary. Overall, the lack of a simple, singular outcome from these tests seems to support the contention that the specifics of time, place and circumstances overwhelm any general pattern.

The association of winning with low slopes and the lack of a significant relationship between the duration of wars and slope suggests that, although guerrillas can survive for a long time in a variety of landscapes, when it comes to seizing power, they must come out of the hills and woods and contest for the heavily populated flatlands that are usually the national core.

CONCLUSION

What tips the balance of numbers in war is the will and ruthlessness of the protagonists. In the early 1940s Japan easily contained the 400,000 Red Chinese guerrillas with 250,000 men. On the other hand, a few people employing terror as a weapon can cause a lot of grief—as the IRA has demonstrated for twenty years. The balance in these affairs is more than purely one of destructive power. There was a current of opinion that the British Army could eradicate the IRA, if they acted as ruthlessly as the Red Army did in the 1920s when it extirpated the Whites using Tukhachevski's tactics. In the 1920s in Ireland, however, the British military experienced the reaction that stepping up the intensity and violence of counterinsurgent operations recruited directly for the opposition. The Soviets obviously pondered this problem in determining to withdraw their imperial reach in Asia and Europe.

Clearly the external circumstances, local details and chance that come into play in determining the final outcomes of violent conflicts make any attempt at formulation and quantification highly conjectural. Nevertheless, a leader of guerrillas as perceptive as Lawrence of Arabia (1957) wrote "the algebraical factors are in the end decisive." In this he was referring to factors calculable in similar terms to those presented here. Thus, it is worthwhile pursuing the issue along these lines to discover the boundary conditions on the balance of superiority at least.

REFERENCES

Asprey, R. B. 1975. *War in the Shadows: The Guerrilla in History*. 2 vols. New York: Doubleday.

Griffith, S. B. (trans.) 1963. *SunTzu: The Art of War*. London: Oxford University Press.

Laqueur, W. 1976. *Guerrilla: A Historical and Critical Study*. Boston: Little, Brown and Company.

Lawrence, T. E. 1957. "Guerrilla Warfare," in *Encyclopaedia Britannica*, vol. 10.

Liddell Hart, B. 1967. *Strategy*. New York: Praeger.

Rezendes, S., and P. O'Sullivan. 1988. "Terrain and the Effectiveness of Guerrillas." *Papers and Proceedings of Applied Geography Conferences* 9: 234–39.

10

WAR IN CITIES

The stand down of NATO and Warsaw Pact forces in Europe obviates the need for the two major military establishments in the world to be prepared to refight the battles of 1944 and 1945 on the temperate plains of Europe. Given the age, fertility and affluence of the post-industrial societies of Europe and North America, it seems unlikely that war will break out there. On the other hand, a full-blown war involving national or ethnic territorial rivalry is very possible, especially in southern and eastern Europe. Obviously riot, suppression and terrorism are likely to persist throughout Europe. However, the most probable venues for fighting with civilian and combatant deaths in the thousands are in the regions of young, growing populations involved in keen competition for power, land and resources. In the past such wars between power elites and classes or ethnic factions largely took place in the countryside, where most people dwelt. However, since it has always been necessary to seize the cities to change a government or a system, towns cannot be avoided. These are the prizes of war—the symbolic and real foci of power.

In the latter part of the twentieth century, the underemployed population of peasant societies continues to flock from the countryside to the uncontrolled and unserviced fringes of the towns to survive. The focus of discontent is shifting to an urban setting. Although the desires of the rural poor remain unrequited, the number and concentration of those who have left the land, as well as their volatility and their opportunity for organization, increase apace. Residence in the capital in particular is liable to generate agitation, since this is where the disparities in well-being are most striking.

If we look at the whole spectrum of violent competition, then at the extreme of nuclear war, towns are clearly not defensible, Star Wars dreams notwithstanding. If we treat cities as prizes to be won, rather than as targets for nuclear spite, there is a limit to the violence that can be done and still maintain the urban fabric, and there is a potential balance to be struck between offensive and defensive advantage. Whether we are dealing with insurrection, civil war or war between nations, the buildings and places that must be seized to win a war or to overthrow the existing order are set amidst large expanses of urban landscape. Either conventional or revolutionary warfare increasingly will involve fighting through built-up terrain. The received wisdom has it that the urban setting, which lends itself to the defense, magnifies the usual 3:1 ratio necessary to overcome a dug-in defense in open country with an armored attack to 5:1 or more. This is an unproved proposition, and it is possible to construe the advantages of accessibility and cover and concealment presented by the streets and buildings of a typical townscape as accruing to the offense at least as much as to the defense. It is a reasonable conjecture that the modern city is easy to take and hard to hold. The political implication of this would be that military domination is an insufficient condition for governmental stability. As long as social discontent is widespread, the prospect always exists of a revolutionary force overthrowing the government by capturing the capital.

Cities then loom larger and larger as the crucial elements of the military landscape. Towns are the foci of power and wealth and the dwelling places of an increasing share of the population. At this time, just over 40 percent of the world's population lives in settlements of more than 20,000 people. Early in the next century, half of the people in the world will live in towns. Towns have always invited the attention of conquerors or spoilers. They are in essence a transport phenomenon. Buildings housing people, goods, machinery, institutions and information are clustered together to reduce the friction of distance entailed in production, exchange, governance and household life. The advantages of close proximity are traded off against the need for space. Territory is controlled economically and politically from its urban nodes. Towns, then, are the primary targets of war. In societies where over half of the population is urban, it is essential to conquer the towns to seize control of the government. Even in the predominantly rural societies, towns are of overwhelming importance in that they articulate the supplies of goods and the power of command. The geographical setting for war will increasingly be the artifice of the townscape.

If we are considering insurrection or civil war, the obvious revolutionary strategy is to drive for the capital as soon as possible. Rebels can survive with a lot of territory at their command in remote, rural regions, but they cannot achieve sovereignty over a nation until they seize its core. In August 1988, Abimael Guzman Reynoso, founder of the Sendero Luminosa, told a *Newsweek* correspondent that they were ready for the "final assault"—taking the cities (Contreras 1988). Although, evidently, they were not actually ready, the sentiment showed a clear recognition of the necessity.

For those in power, it makes sense to avoid being drawn into peripheral operations at a cost of thinning power in the core area. Since rebel forces usually have the advantage of greater mobility, there is the prospect that they can create space to attack the capital by drawing government forces out onto the fringe and then by filtering through government lines to reconcentrate near the capital against depleted opposition. If the government can withstand the embarrassment of "no-go" areas for its writ within its frontiers, it can survive as long as it dominates the economic and demographic core area with the confidence of most of the people there.

URBAN TERRAIN

The modern town, be it in the First, Second or Third World, is not essentially a defensive structure, like its ancient and medieval predecessors. The age of defensive urban design ended with Vauban, who wrote the text on destroying earthwork defenses in depth against artillery, even as he was laying out the finest examples of this art form. The towns that have grown up since 1700 have been laid out with functional or symbolic rather than defensive objectives in mind. For most of the urban ground cover of the globe, the chief design criterion and organizing principle has been efficient production of goods and services via the enhancement of mobility. Towns are theorized about in terms of minimizing the cost of friction to a mass of population and activities in a setting where places of work require enormous numbers and close proximity to each other, so that a daily burden of travel between home and work place is necessary. The urban fabric has been constructed largely to minimize the cost of the daily tidal flow of people into and out of massively concentrated places of work and business. Over this, especially in capital cities, has been laid the symbolic composition of the urban landscape—the concrete expression of power and dominance. Engineering and architecture have been used to focus attention on the state and commerce, expressing the overwhelming obsession of the modern world with politics and money. The grand avenues, squares and systematic gridding of central cities converge on the offices of state and the seats of economic and political power.

If there was a military objective in urban design, it was to make the hearts of capitals indefensible to prevent the mob from seizing and defending the center of the state. So the narrow, baffled streets of the medieval core of Paris were replaced in the 1860s by Baron Haussmann's boulevards, accessible to light, fresh air and infantry. The rookeries, the lair of the London mob in St. Giles, Drury Lane, Clare Market and Saffron Hill, were broken up with bridgeheads, avenues and metropolitan improvements in the last half of the nineteenth century. The layout of Soweto, the black township on the margin of Johannesburg, lends itself to armored forays from the fortified police stations which are its chief public buildings. The military rulers of Myanmar instituted an "urban renewal" of Yangon at the start of 1990 to break up the neighborhoods of squatters who had supported the prodemocracy movement.

We have noted that the Paris Commune of 1871 was put down in a relatively short time. The Army of Versailles easily circumvented any barricades the communards erected. The communards did not understand the effects of Haussmann's remodelling of Paris on the techniques of street fighting. This was not the city of 1848. By the 1870s it was obvious that the cities were soft targets for revolution; however, although it was easy to grab control, it was difficult to harden the defenses sufficiently to resist a counterattack. The only possibility of survival was a destructive last stand, retreating from house to house and burning what could not be held. Dublin in 1916, St. Petersburg in 1906 and again in the spring of 1917 showed how difficult it was for revolutionaries to hold capitals. By October 1917 it was the wholesale conversion of soldiers and sailors to the Bolshevik camp that ensured Lenin's success, not any fight for the city. The Whites' only real battle was the last stand in the Winter Place. The lessons of history were sufficiently clear to the IRA by 1972 that, when they learned about the British build up for Operation Motorman, designed to mop them up in urban no-go areas, the Provo command determined that they were not equipped for a static defense against an armored force. The IRA decided to avoid fighting in a confined area against superior numbers and equipment and melted away on July 3 as the British Army advanced into the Bogside, the Falls, Portadown, Lurgan and Armagh pushing aside unmanned barricades.

TACTICAL BALANCE

Turning from the past, for the foreseeable future, it seems likely that politically crucial fighting will take place in a city setting. This is where power will be lost and won—in Kabul, Jammu, Manila, Beirut, Luanda, Kuwait, Addis Ababa, Khartoum, San Salvador and Lima. This raises two purely objective, geographical questions. Does the urban landscape favor aggressors or defenders? What tactics are likely to be most successful in either pursuit? The answers to these questions, then, provide food for political discourse on the likelihood of success of the insurgents versus the incumbent governments and, thus, on political and social stability.

When we turn to military doctrine, of the U.S. Army for example, we find that the townscape is judged to offer the advantage to the defender. This doctrinal position is, however, couched in the central European context and is concerned with conventional, armored operations. The question can be more precisely posed in terms of the balance of force. Received wisdom has it that it takes at least three times the aggressive force to overcome any dug-in defensive force in open countryside. Our question then becomes whether the variety of urban terrain brings about a departure from this ratio in either direction. This is largely a matter of the ease of access and movement presented by the terrain, the benefit of which accrues to the attacker, versus the characteristic sight lines of the setting. The longer the sight lines, the better the fields of fire that can be established, the easier it is to defend a certain frontage. The bigger the frontage a basic unit of soldiers can hold, the larger the perimeter that can be defended with a given number of soldiers and weapons.

Taking mobility first, let us consider an idealized urban landscape, a square grid network of roads. Given that aggressors have a considerable advantage in speed, since they do not have to move emplaced weapons, they should be able to amass the 3:1 advantage in firepower necessary to break through along certain avenues readily or to find an undefended avenue of approach, if we allow a hesitation by the defenders, before they realize what is happening, of the collective equivalent of the five seconds' advantage of surprise in individual combat. The rectangular lattice is not a mere abstraction. It is a widespread feature of the urban landscape, especially in city centers. When we consider sight lines, the orthogonal grid does provide long vistas in its cardinal directions, but not much otherwise. For oblique fire, the line of sight will depend on building setbacks and street widths, which may not be enormous in the closed blocks of the downtown area. If blocks are an eighth of a mile long, then the effective maximum line of fire may be only 220 yards, since tanks can hide and dodge out of side streets. Given that a tank can get off a deadly round more rapidly than any antitank weapon, the downtown may not be all that healthy for the defense. After their initial success against tanks without infantry support, the Hungarians succumbed in just over a week to a more balanced, well-prepared Soviet tank attack on Budapest in November 1956.

Sight Lines in Cities

Although the military implications of the urban landscape have been debated, little has been done to measure the fabric of towns from the viewpoint of fighting with modern weapons. If we take official U.S. Army doctrine, for example, enshrined in *FM 90–10, Military Operations in Urbanized Terrain*, we find specific tactics and techniques for offensive and defensive deployment in a series of categories of townscapes. The tactical guidance offered is based on assumptions about the geography of buildings, streets and open spaces in the landscape. John Quick (1988) made some measurements of the most significant military aspects of the urban terrain of two fairly typical German towns which called into question the validity of the implicit assumptions of U.S. Army doctrine, especially with regard to the extensive and growing expanse of suburbia.

The field manual establishes five classes of building and street patterns:

Table 10.1
Fighting Sector Widths

	Attack zone width in meters	Defensive sector width in meters
A	175	400
B	250	500
C	350	650
D	400	700
E	500	900

Source: FM 90–10.

A Dense, random construction
B Closed, orderly blocks
C Dispersed residential areas
D High-rise areas
E Industrial, transportation areas.

The order of presentation of these implies a simple trend of increasing openness between the classes. The underlying assumption seems to be that building density decreases, and, thus, the line of sight distance increases, as one progresses from A at the center of the town, through B to C and D to E on the periphery. That this is the implicit image of urban terrain can be inferred from the typical median attack zone and defensive sector widths that are prescribed for various patterns.

The monotonic increase of these values between the classes has all the appearance of a crude linear extrapolation, rather than the result of measurements (see Table 10.1).

These assumptions about urban geography, then, provide the basis for evaluating each type of terrain for combat. Defense and assault are considered with respect to the mobility, observation lines, fields of fire, obstacles, cover, concealment and limitations on command and control inherent in each category of townscape. The assumed density gradient among types of terrain translates into generalizations on the allocation of troops and weapons. The force required to attack or defend successfully is deemed to diminish as you move from A to B to C to D to E. No empirical support is offered to support this contention.

To test the validity of this assumed simple progression of openness, some measurements were made on two German towns. Schweinfurt and Bad Hersfeld are covered with unclassified aerial photography at a reasonable scale for analysis, and both contain a significant extent of all five of the designated building patterns.

FM 90–10 describes these patterns as follows:

Table 10.2
Line-of-Sight Distances

	Terrain category	Percent of total area	Number of measurements	Mean line of sight distance (meters)
	A	8	11	85
	B	11	15	240
Schweinfurt	C	37	49	173
	D	14	18	101
	E	30	40	387
	A	7	9	47
	B	2	4	185
Bad Hersfeld	C	59	78	115
	D	5	6	140
	E	27	36	268

Source: Quick and O'Sullivan, 1988.

A. Old inner city construction with narrow winding streets radiating in an irregular manner from a central area; buildings located close together and frequently along the edge of the roadway.

B. Wider streets forming a generally rectangular pattern with buildings frequently forming a continuous front along a block.

C. Row houses or single dwellings with yards, gardens, trees and fences; street pattern regular or curving.

D. Multistoried apartments separated by large open areas and one story buildings; wide streets laid out in rectangular patterns.

E. Low flat-roofed factories and warehouses; located on or along major rail and road routes.

Bad Hersfeld and Schweinfurt were divided into these categories of urban terrain. The percentage of each town in each terrain category is given in Table 10.2.

A sample of points for measurement was laid down on the aerial photographs and distributed among terrain types according to these proportions. The measurement that coincides most closely with the intent of the FM 90–10 classification is line-of-sight distance. This has a critical impact on the range of observation, the fields of fire and tactical communications. Indirectly, it affects concealment and ease of command and control. This, then, was used to compare with the attack and defensive zone widths specified in the manual. Line-of-sight distances were measured from the sample points, observing 360 degrees around each point and measuring the maximum distance before obstruction. The average distances for the terrain categories are given in Table 10.2.

It is evident from these results that the progression of openness implied

Figure 10.1
Tactical Widths and Line-of-Sight Distances

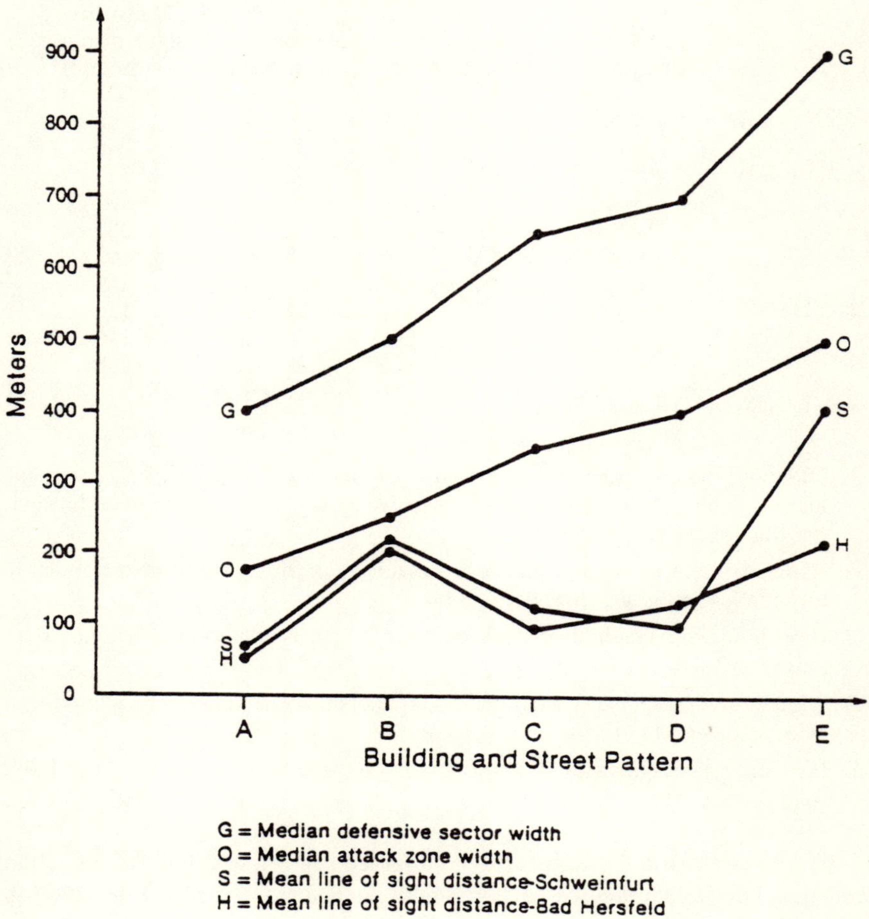

in FM 90–10 does not hold in these cases, which are not atypical of urban Europe. There is a disparity between measurements of maximum lines of sight and the sector widths of doctrine. This is illustrated in Figure 10.1. In C and D areas, typical line-of-sight distances are much lower than the doctrine would lead us to expect. This departure takes on greater significance when it is noted that these geographical patterns account for over half of the real estate. In Schweinfurt, types C and D account for 51 percent of the built-up area; in Bad Hersfeld, they cover 64 percent of the ground.

It seems that the degree of road curvature and irregularity in alignment in C and D areas offset the greater building setbacks and the amount of open space. Maximum sight lines are shorter than those that traverse the

Table 10.3
Line-of-Sight Distributions for Dispersed Residential Areas (C)

Line of Sight in meters	Cumulated Percentage	
	Schweinfurt	Bad Hersfeld
0-50	19	25
51-100	23	27
101-150	16	14
151-200	10	19
201-250	6	10
251-300	8	4
301-350	0	0
351-400	10	1
401-450	2	
451-500	2	
501-550	4	

Source: Quick and O'Sullivan, 1988.

regular blocks of B type areas or cross the wide open vistas of the industrial landscape of E. Departures from a regular grid arrangement of streets, in particular, cuts down the sight lines in C areas significantly.

What does this signify in terms of fighting? Suppose, for argument's sake, we have forces employing American tactics defending a town like Schweinfurt against an armored attack in the Soviet fashion on the basis of the prescription for sector widths given in FM 90–10. More than half of the town and over two-thirds of the outer edges of the town are of types C and D. Thus, the likelihood is great that fighting will start in such a setting. The evidence suggests that a one-company combat group would be allocated by the aggressor to attack down each street. In dispersed residential areas, this could mean one company every seventy-five meters. Given the sector widths in FM 90–10, this would give the attacker a combat ratio advantage of 9:1.

Now for the tanks. According to U.S. doctrine, tank mobility is generally good in suburbia, given big housing setbacks. How can tanks be knocked off in this setting? The more powerful American antitank weapons, the TOW and the Dragon, have minimum line-of-sight ranges for firing in the open of 110 and 95 meters, respectively. Table 10.3 shows the distribution of maximum lines of sight in Schweinfurt and Bad Hersfeld. In Bad Hersfeld, 60 percent of the measurements of *maximum* lines of sight in dispersed residential areas were less than the required 110 meters; in Schweinfurt, 52 percent of the measurements were below this range.

The problem of finding a suitable position from which to kill tanks is greater than this. Apart from the 30- or 40-meter back blast clearance required, it is necessary to have a vertical clearance of over ten feet over any obstacle that might interfere with the flight of the missile. Fences, parked cars and bushes, all in abundance in suburbia, would further restrict firing lines. With these weapons the gunner must guide the missile until it strikes. If the enemy tank moves behind a building or any other screen, the gunner is out of luck. Even if the tank stays in the open, at ranges between 100 and 175 meters, the TOW is rated as having only a 70-percent chance of hitting a target and the Dragon a 55-percent chance. If the attacking tanks are as good as American tanks, they have a 95-percent probability of hitting their mark with the first round. So the advantage would seem to lie with the attacker. The light antitank weapon (LAW), with its minimum range of ten meters, could be useful in this setting, but it would have to be used in volleys to stop a tank. The best tank killer is another tank. C and D type terrain do present some problems for their defensive deployment, however. Trying to hide a thirty-foot-long by eleven-foot-wide chunk of metal and machinery is not easy in suburbia. Even the most spacious of front rooms might be insufficient, especially if they have cellars beneath them and floors not designed to bear a fifty-ton load. The M1 tank's range finder has a minimum reading of 200 meters. In Schweinfurt, only one-third of the *maximum* line-of-sight distances are greater than this and in Bad Hersfeld only one-sixth.

Command and control in suburbia may present more difficulty than anticipated. A company commander's visual field will be far smaller than FM 90–10 implies. Radio transmissions may also suffer more than foreseen.

Line-of-sight distances affect the ability to see the enemy and to lay down fields of fire. Thus, they affect the selection and use of weapons and the control that can be exerted over operations. The results of these measurements suggest that current U.S. doctrine is defective in assessing line-of-sight distances and combat conditions in a large and growing part of the modern landscape.

Mobility

Although cities whose main growth has been in the era of the internal combustion engine have fairly uniform road densities, somewhat older cities, which grew with mass transit, have road networks that are tighter near the center. The frequency of intersections diminishes from the downtown toward the suburban fringes. If we set aside the variation in peak traffic levels, the urban velocity field increases from the rural edge and peaks at the city center where road surface exceeds any other use of land. It is only in consideration of the congested network that achievable speeds diminish toward the center of town. If we assume that civilian traffic would clear the streets in the event of a battle, then the potential time taken for a random walk of given length will decrease with network density right to the middle of town. This advantage to the offensive is potentially offset by a greater mass and height of the buildings, which provide cover and concealment. Concealment can, however, be friendly to the attack, if it can bring lethal fire to bear more rapidly than the defense. The inner city, with its intense network of underground routes and service tunnels, does provide cover and concealment below the surface, which is primarily to the advantage of the defender. These observations apply to towns of the developed world; however, the crucial wars of the future will likely occur in Third World cities. The obvious question arising is whether these are different in structure in ways that tip the tactical balance.

THIRD WORLD CITIES

Few measurements are available on the relevant characteristics of the cities of Africa, Asia and Latin America. Although similarities to European and North American cities are discernible, there are some notable distinctions. The simplest and most elegant measure of urban structure is the population density gradient which describes the variation of residential population density with distance from the city center. The cities of the industrialized world in general display a negative exponential relationship, with a tendency through time for this to flatten with a reduction in distance friction and urban growth. The limited calculations done for Third World cities show similar tendencies. Cross-sectional evidence shows that gradients diminish with greater total population. The few observations through time for one city, show a tendency to flatten with growth. There are, however, departures which reflect historical circumstances. The white cantonments or lines close to the center of colonial south Asian and African cities and suburbs in privileged central locations reduce the central density below that of traditional cities of those regions. Latin American cities have gridded centers prescribed by the Laws of the Indies. They also display a high concentration of the upper classes in residence near their centers, reflecting tradition, lower car ownership and poorer public transportation. The poor fleeing the countryside cannot gain admission to the established city and so share the unserviced periphery or awkward terrain with large-scale and objectionable land users, such as airports, military establishments, industrial plants and rubbish dumps. Latin American *barriadas* or *flavelas* have their counterpart in the shantytown fringes of African cities. The flatter gradients produced by this irregular pattern of growth on the outer margins, engulfing old villages into a structure geared to very locally circumscribed objectives and needs, is characteristic not only of Latin America but also of Asia and Africa. One remarkable feature of Latin American cities is the spine-like extension of the business district and elite residences from the center along a principal radial. The Paseo de la Reforma of Mexico City is the classic example of this phenomenon.

None of these features holds out additional comfort for the defense. The spinal development in particular represents a geographical hostage against the use of air and artillery superiority by the incumbent. Oligarchies are reluctant to destroy their constituency in order to protect them from political change. The spine then becomes a favored avenue of approach.

The central locations of the upper class restrained U.S. Air Force assistance to the Aquino government against insurgents in Manila in December 1989. F4 Phantoms merely showed off air superiority to express U.S. support for Mrs. Aquino against the 6,000 rebels who so readily seized the center of the city. Although this was a demonstration by the U.S. of the immediate political potency of armed might, it did not establish a secure political foundation for stable government. The same factors gave rebels an advantage in San Salvador in November 1989, where government forces were reluctant to use artillery and command of the air to dislodge them from downtown. The unrestrained U.S. airborne attack on Panama City in December 1989 showed the civilian cost of heavy firepower brought to bear in the central city.

Both theoretical abstraction and recent events, then, suggest that Third World towns are as vulnerable to attackers and are as difficult to defend as are, theoretically, the cities of the First World. If this is so, then military superiority is an insufficient basis for stable government; the hearts and minds of the people remain to be conquered.

REFERENCES

Contreras, J. 1988. "Shining Path." *Newsweek* (August 15): 32.

Griffin, E., and L. Ford. 1980. "A Model of Latin American City Structure". *Geographical Review* 70:397–422.

Lowder, S. 1986. *Inside Third World Cities.* London: Croom Helm.

O'Sullivan, P. 1980. "Warfare in Suburbia." *Professional Geographer* 32(3): 355–60.

Quick, J., and P. O'Sullivan. 1988. "Military Analysis of Urban Terrain". *Professional Geographer* 38(3): 286–90.

Trainor, E. 1990. "El Salvador: The Battle for the Cities." *Marine Corps Gazette* (January): 8–11.

U.S. Department of the Army. 1977. *The Tank and Mechanized Infantry Company Team.* Washington, D.C.: U.S. Government Printing Office. Field Manual 71–1.

———. 1979. *Urbanized Terrain (MOUT).* Washington, D.C.: U.S. Government Printing Office. Field Manual 90–10.

Wheatcroft, A. 1983. *The World Atlas of Revolutions.* New York: Simon and Schuster.

11

NORTHERN IRELAND

The best way to integrate the notions developed in the last several chapters is to draw their strands together in a sustained example of the interaction of geopolitics, strategy, tactics and terrain. Although it does not qualify as a war in Ruth Sivard's terms of causing over 1,000 deaths a year, the conflict in Northern Ireland is seen by the combatants as a revolutionary war; it has been waged in both the countryside and towns. Through its twenty-year duration, there has been an evolution of strategy and tactics reflecting geopolitical stances and the realities of the physical and human terrain. To reduce the complexities involved to manageable proportions in the interest of clarity, attention will be focused on the policies and actions of the British Army and the IRA, the chief protagonists in the armed struggle both of which have explicitly violent missions. The doings of the Royal Ulster Constabulary (RUC), protestant paramilitary groups and republican splinter groups will be left aside for the most part.

HISTORICAL SETTING

In 1922 the six northeastern counties of Ireland were established as Northern Ireland with a parliament in Stormont and a prime minister and cabinet inevitably drawn from the Unionist Party, which represented the coalition of interests of the Protestant majority. The Nationalist Party, representing the Catholic third of the population, was gerrymandered out of even one-third of the seats in parliament and was doomed to perpetual minority status.

In 1956 the IRA, the minuscule remnant of the guerrilla group that had made life too uncomfortable for the British government in the rest of Ireland during the early 1920s, launched a terrorist campaign, blowing up government, military and police facilities. In six years, six members of the RUC and nine members of the IRA were killed. In the end the IRA gave up, exasperated at the lack of Catholic response to their call to rise up and throw off the Unionist yoke. By the early 1960s, the Catholic population of Northern Ireland were more vexed by rising unemployment and discrimination against them over jobs and housing.

In the 1960s Captain Terence O'Neill, the new Unionist prime minister, adopted a more paternalistic and conciliatory attitude toward Catholics. This aristocratic stance brought an extreme Protestant reaction, led by the Reverend Dr. Ian Kyle Paisley. This found a violent expression in the Ulster Volunteer Force, which terrorized any Catholics suspected of IRA sympathies. O'Neill's liberalism was not manifest in action. Although they accounted for more than 30 percent of the population, Catholics got only 12 percent of the local government jobs and 6 percent of the civil service positions. Given a large number of Catholics with university educations, this caused discontent among an emergent Catholic bourgeoisie, who began to express themselves through the Northern Ireland Civil Rights Association, whose campaign of marches began in 1968.

The beginning of the disorder which still prevails can be traced to a march in Derry in 1968. The participants avoided Catholic, nationalist slogans and stuck to demands for civil rights. The marchers were middle-class teachers, students and a Westminster member of parliament, Gerry Fitt. The 2,000 marchers were met by an RUC baton charge, which left seventy-seven wounded civilians, including Fitt. The civil rights movement suspended its activities to give the government time to sort things out, but a radical group of students staged another march in January 1969. This march was ambushed by loyalist activists. In the spring of 1969 there was an outbreak of bombing Catholic churches.

In August 1969 the Orange Order's Apprentice Boys of Londonderry held their annual parade from the old, walled city past the edge of the Catholic Bogside. Their customary anti-Catholic jeers and penny throwing were met this year with a hail of stones. The Apprentice Boys retaliated and the RUC prepared for action. The Bogsiders threw up barricades and waited for the RUC who charged in at 7:15 P.M. The news of the Battle of the Bogside spread to the Catholic areas of Belfast, and crowds stoned police stations on the Falls Road. Mobs of Catholics and loyalists were soon doing battle in the streets. A loyalist crowd marched into the Catholic enclave but was stopped when one of its members was shot. This was the signal for the RUC to roar into the Falls in armored cars with Brownings blazing. There were riots in Armagh, Dungannon and Dungiven as well. On August 14 the British Army was ordered in to maintain law and order. The IRA was nowhere to be seen because its handful of members had been surprised by the turn of events.

In this book I have employed a division of violent conflict into tactical, strategic and geopolitical levels of decision, according to the scope of the issues involved. Tactics is concerned with the deployment of people and weapons in the field; strategy, with the management of war for given political ends; and geopolitics, with the determination of the ends. Although this book focuses on tactics and treats them as a matter for cold, military calculation; in reality, the style of fighting adopted usually reflects the self-perception, self-deceptions and doctrinal positions of the parties to the affray. Thus, it makes sense to start with the geopolitical positions of the opponents in this twenty-year-old battle and to trace the interactions of these with strategy and tactics.

GEOPOLITICAL OBJECTIVES

To start with, then, it is appropriate to consider the avowed and implicit objectives of the British government and the IRA and to establish the agenda for the conflict.

The British

Whatever the fears and ambitions of the Protestant population of Northern Ireland, captured in the phrase "siege mentality" and the slogan "a Protestant Parliament for a Protestant People," the attitude of the British establishment to the region is largely a matter of geopolitics and the containment of class turmoil. Whether or not it is sincere, the British government has made a fuss about Irish neutrality and its implications for command of the sea passage lying between Greenland and Iceland and Ireland and Scotland. In the 1980s the "GIUK Gap" achieved greater significance as one of the choke points identified with the more aggressive stance of the navies of NATO in their global competition with the USSR. A 1949 British cabinet memo [PRO C(49)4] placed such importance on the role of Northern Ireland in British defense that "it should not be allowed to leave the UK, even if that was the desire of the majority of the population there." Representing the viewpoint of the Republic's establishment, Charles Haughey evidently sees the strategic issue as the foremost one that exercises the British government. It is noteworthy that Enoch Powell, who led the Ulster Unionists of British nationalist bent in seeking full integration of Northern Ireland into the United Kingdom, publicized a "Foreign Office conspiracy" directed against such a union. This conspiracy theory had it that the Western intelligence services were misinforming their masters so as to bring about the inclusion of the Republic of Ireland in NATO. The obvious implication is that if this were done, a political solution to the local problem of Northern Ireland involving the Republic could be achieved without compromising the integrity of command of the entrance to the North Atlantic. The strategic significance of the GIUK Gap and, thus, of Irish neutrality may have evaporated, but the British establishment takes some time to respond to events.

The British Army was brought into the Northern Ireland conflict "in aid of the civil power." There are grounds for suspicion, however, that some members of its officer corps did have a mission of their own which could put them at odds with politically expedient solutions to the conflict. The precedent was established in 1914 when the chief of the imperial general staff gave a written assurance to the commander of the cavalry brigade in Kildare that his troops would not be called upon to enforce the Home Rule Bill in Ulster. More recently, writing on the use of soldiers to maintain law and order, the chief British counterinsurgency strategist, Frank Kitson, who retired from command of United Kingdom land forces in 1985, pointed out that the standing army was first raised in the seventeenth century for the "Suppression of the Irish" and the "Defence of the Protestant Religion" (Kitson 1971, 24). Clearly, there is an implicit notion of the continuity of a cultural imperative that transcends the authority of mere politicians. There is more than a hint in Kitson's writing that soldiers may find it necessary to contravene the civil authority to suppress insurrection and to maintain the established order as they understand it. "Subversion in a good cause may be the right thing for a soldier to do" (Kitson 1971, 8–9). Although he does not quote any sources in his work, given his experience in Malaya, Kenya, Cyprus and Oman and his stint as a fellow in University College Oxford in 1960–70, Kitson must have been acquainted with the works of the French apostles of *la guerre revolutionnaire*. These veterans of Indochina and Algeria portrayed guerrillas who drew power from their Marxist inspiration. Like the French, Kitson prescribed the use of psychological warfare, clandestine methods and the indoctrination of soldiers and civilians in the fight against the global spread of communism. It appears as if the British military establishment feared more widespread, internal disorder in the United Kingdom and a danger to the existing class structure, especially from the immigrant population. Kitson's writing and his posting to command the British forces in Belfast for two crucial years, from September 1970 to April 1972, provided a laboratory and training ground for antirevolutionary units and tactics with "real" soldiering against a "genuine" enemy. In 1984 Sir John Briggs Davidson, the Tory spokesman on Northern Ireland, justified the possibly lethal use of rubber bullets against rioters by citing Marighela and the danger of Marxist revolution.

The IRA

The leadership of the Provos (the Provisional Army Council of the Irish Republic Army), who were responsible for most of the violent attacks on the status quo, were not communists, but apolitical revolutionaries. They saw themselves standing in an unbroken tradition leading back to Wolfe Tone and the men of 1798. Although the Official IRA, the "Stickies," and some splinter groups have veered to the left, the Provos remained republican nationalists who believed that all of the institutions of the six and twenty-six counties of Ireland were illegitimate and must be overturned to create a united republic by force of arms. Playing politics was merely a means to this end.

STRATEGY

The power gradient model of insurrection discussed in Chapter 5 suggests rational strategies for government and for insurgents. Whether or not these were conceived of and followed can, on the one hand, be inferred from observations of the geography of warlike behavior. On the other hand, we can turn to the published or disclosed plans, statements and decisions of the leadership of the contestants, to see if they accord with our expectations. Both of these means will be employed here.

First we consider the manifestation of strategy in the spatial pattern of violence. Our theoretical model would predict that, if both parties were strategically rational, we could expect to see a concentration of violent confrontations in peripheral areas in which the insurgents manage to maintain the upper hand. This would conform with the intersection of the power cones in our simple geometric model. If the insurgents were serious about overturning the government, they would have to attack the government presence in the capital at every possible opportunity.

Looking at the distribution of violent incidents involving the British Army in 1974, when the shooting war was three years old and past its peak but before it reached a stalemate, we should see a clear reflection of the strategic stance of both sides. The distribution (Figure 11.1) bears a strong resemblance to our theoretical expectation with concentrations in south county Armagh and around Strabane and Derry, with the heaviest clump in Belfast. In Belfast we find a microcosm of the regional picture with a concentration in the Falls and Ballymurphy and the downtown.

There is, of course, another historical interpretation of the geography of violence. There is a strong correlation between the level of violence and the proportion of Catholics in the population of an area. The historical explanation for this disposition traces from the plantation of Ulster in the sixteenth and seventeenth centuries. This we could interpret as a human terrain factor s reinforcing the strategic inclination. It is here where the support for the IRA is the deepest. The rural areas of Catholic preponderance not only are near the border but also are the least fertile and the most rugged with plenty of concealment. The urban areas occupied by Catholics are densely packed with houses and provide ample screening and cover to the west of downtown Belfast and Derry.

When we turn to the conscious expressions of strategy, we may be able to separate out the intentions from the happenstance of war. This discussion of strategy and the following treatment of tactics will be confined to the

Figure 11.1
Incidents Involving the British Army in 1974

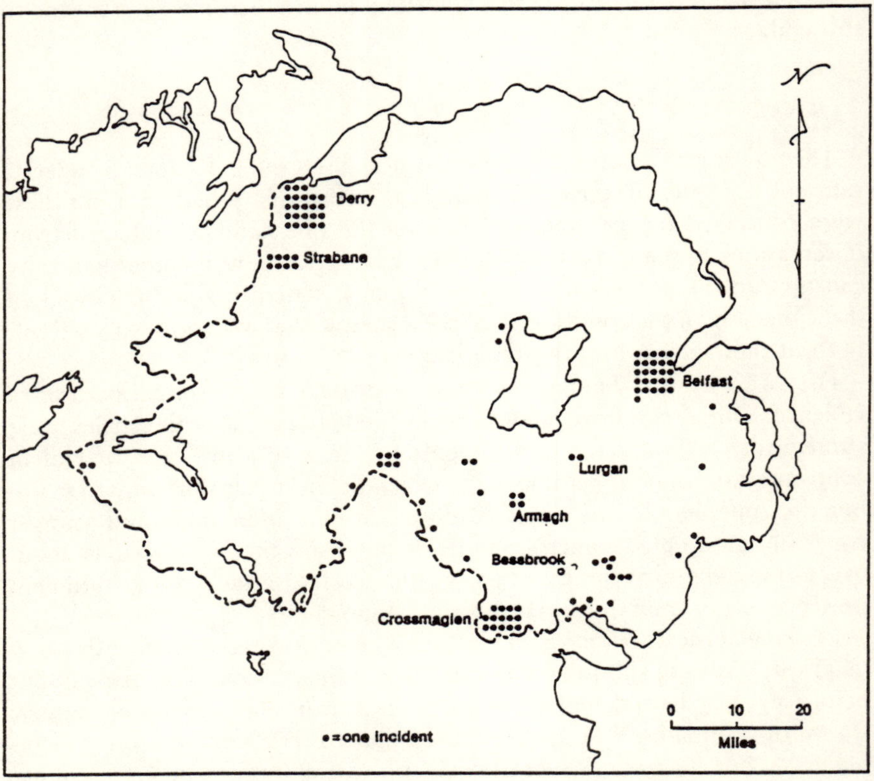

activities of the British Army and the Provos, concentrating on the period between 1970 and 1980, when the battle was most fiercely joined. Since the British Army was in the field first, we will begin with what we can glean of their stratagems and then turn to the IRA.

British Strategy

In 1969 British troops were not sent into urban neighborhoods to protect Catholics from loyalist mobs and rioting police, but to maintain law and order. This turned out mostly to involve protecting Catholics in working-class areas. At this juncture the vestige of the IRA in the Falls Road area had six guns and could do little. The graffiti read "IRA = I ran away." It has been estimated that members of the IRA amounted to 1 percent of the total population of Northern Ireland; however, these few were active in the citizens' defense committees which organized what protection there was, and their membership began to grow.

The "softly, softly" policy of the British Army sought to keep the warring factions apart, to reinforce and legitimize the barriers between Catholic and Protestant communities and to establish a "peace line" along Divis Street, which separated the Protestant Shankill from the Catholic Falls. Behind these lines, the "no-go" areas developed. Although the British Army patrolled them and fraternized with the locals, the RUC were excluded. It was in this setting, in the winter of 1969, that the Provos were established and took on a defensive role.

The "honeymoon" between soldiers and Catholics was disturbing to the loyalists, the IRA, and the British Army command. In a meeting of the Stormont cabinet on August 14, 1969, Chief of Staff Brigadier Dyall said, "[I]f the police, B Specials and Army were totally committed, then we could beat the Micks" (Hamill 1985, 13). General Officer Commanding (GOC) Sir Ian Freeland predicted that "the honeymoon between troops and local people is likely to be short lived" (13). Although the Labour government under Wilson and Callaghan had already adopted a pro–status quo and, thus, pro-Unionist stance in early 1970, the radical shift came with a Tory electoral victory in June 1970. In a press briefing, Defence Secretary Lord Carrington let it be known that the British Army could not remain impartial and fight on two fronts but would rather seek to suppress the Provos by any means available and hope that loyalists would remain quiet. In September 1970 Brigadier Frank Kitson was given command of the 39th Brigade and, thus, of Belfast. In 1975 Paddy Devlin, who was to be a member of the ill-fated Northern Ireland Executive, wrote that "it is clear from public records that his [Kitson's] troops were instructed to put the boot into the catholic sections of the community" (Davis 1986, 182). In 1971 Kitson remarked, "I would squeeze the catholic population until they vomit the gunman out of their system" (182). In more scholarly fashion, in *Low Intensity Operations*, he said that "conditions can be made reasonably uncomfortable for the population as a whole, in order to provide an incentive for a return to normal life and to act as a deterrent toward resumption of the campaign" (Kitson 1971, 87). Commenting on Mao's "guerrilla fish swimming in the water of public support," Kitson suggested that the water might have to be "polluted" (87). British soldiers were told that their task was to reassure Protestants and assist Catholics to lose their romantic hope of reunification by force with the Republic.

In July 1970 sweep-and-search operations in Catholic areas along with the curfew imposed on the Falls area stirred things up. Although the IRA members here were Officials and not Provos, and although the area had seen no fighting, it was selected as an experimental "psycho-social target." In a thirty-six-hour period, every house was searched with "maximum irritation and damage." The homes of 20,000 Catholics yielded 200 weapons and, subsequently, many recruits for the Provos.

There was, however, some strategic confusion in the British ranks. There was a tussle between GOC Harry Tuzo and Kitson. Tuzo preferred a colonial garrison policy of using soldiers as a "ring of steel" around dissident areas to provoke an all-out confrontation. Kitson advocated more fluid, infiltrative, intelligence-gathering operations, involving plainclothes, Special Air Services (SAS)-trained, military reconnaissance forces and *agents provocateurs*. Whether or not it was masterminded by Kitson, Operation Demetrius, the internment of 1,500 Catholics mounted in August 1971, was similar to operations he had supported in Kenya for their intelligence-gathering function. This action, which swept up very few active members of the IRA and got few names, again recruited for the Provos.

The growing political inclination was toward containment. In December 1971 Home Secretary Reginald Maulding envisaged a situation in which the IRA would not be defeated but would have their violence reduced to "an acceptable level." British Army operations and, thus, violence were to be localized in the Catholic working-class areas in towns and poorer rural areas to "contain if not extirpate terrorist violence" (Baynard 1985, 125). Unionist politicians were calling to let Derry burn and to concentrate on sorting out Belfast first.

The shooting of thirteen civilians in Derry by paratroopers on January 30, 1972, and the violence that followed brought about the suspension of Stormont and the direct rule of Northern Ireland from Westminster in March. Kitson was dismissed in the same month. The first secretary of state for Northern Ireland, William Whitelaw, initially adopted a less provocative strategy. The British Army were ordered to play the Creggan and Bogside in the lowest possible key compatible with general security. The British Army command felt they could dominate Belfast but not Derry. In the no-go areas, where the writ of the British government was not executed, the Provos ran affairs. Loyalist antagonism to the existence of these enclaves and the buildup of the Provos to 1,500 led to Operation Motorman: 22,000 men were mustered to mount the biggest operation since Suez. The barricades were to be knocked over and the no-go areas reoccupied. It was felt by prime minister Edward Heath that an estimate of 100 likely deaths was a "politically acceptable" level of violence. On July 31, 1972, Sir Harry Tuzo's twenty-seven battalions swung into the Bogside, the Creggan, the Falls, Portadown, Lurgan and Armagh. The Provos faded away and took to the border countryside.

In January 1973, Lt. General Sir Frank King was appointed GOC. He felt that the key to the solution was to remove the IRA from the streets of Belfast from whence the military could dominate the countryside. He thinned out the British forces on the border and in Derry where they were only providing good targets for the Provos. The British Army remained focused on the Catholic front, refusing to confront the Ulster Defense Association and other loyalist groups during their intimidation campaign of 1974. By 1975 the British Army felt they had the upper hand in Belfast. In the countryside, things were otherwise. In that year, nine of twelve soldiers killed were shot in south Armagh by Provo snipers. The British Army, with four companies patrolling 400 square miles, did not have the initiative. The Provos were manning roadblocks. Although the border was not officially a friendly one for the IRA, it was a difficult one for both sides to patrol. The Irish army reckoned it would take two million men to seal it effectively. The GOC redeployed one battalion and 500 men of the Ulster Defence Regiment (UDR) to the border. Early in 1976 the SAS were officially brought into south Armagh. The British Army command talked about pulling out of Crossmaglen back to a border base at Bessbrook. In the view of the commander of the Provos, "The BA would obviously prefer to sit tight in cities and towns hoping to wear down the urban IRA units by a policy of attrition. Instead they were drawn out by the aggressive activity of our units in rural areas" (MacStiofain 1975, 332).

Taking responsibility for Northern Ireland in 1974, Merlyn Rees adopted a reactive policy of containment, leaving the initiative to the IRA. The British Army campaign was to be regulated by the level of intensity of IRA offensive operations. This was clearly a prescription for stalemate. A Ministry of Defence secret intelligence report, authorized by Brigadier James Glover in December 1978, saw no prospect of a Provo demise for at least five years. When this report fell into the hands of the IRA, it was widely publicized. It rated the Provos as professional soldiers with "the dedication and sinews of war to sustain a significant level of violence for the foreseeable future" (Hamill 1985, 239). In 1980 the former head of MI6, Sir Maurice Oldfield (the model for John Le Carre's George Smiley) was appointed to coordinate the British Army and the RUC with a view to reducing the British Army complement to 11,000 with 7,500 UDR. The RUC and UDR would deal with white and gray areas on the map of degrees of hostility, while the regulars would contain the black areas of west Belfast, Derry and the border with heavy surveillance, patrolling and continued efforts at deep penetration by SAS types.

Provo Strategy

The Provos were organized in December 1969 when they broke away from the leftist Officials who did not want to get drawn into the affray. Sean MacStiofain, who had pondered on guerrilla war in H.M. Prison Wormwood Scrubs, while Kitson was doing so in Kenya, organized a series of local, largely autonomous companies for the active defense of Catholic communities. Until the end of 1970, these companies remained defensive, establishing their position in the urban and rural no-go areas with nine units in Belfast, five in Armagh, four in Down, three in Tyrone, two in Fermanagh and two in Derry. MacStiofain saw these as bearing a strong resemblance to Che Guevara's *foco insurreccional*. It was not until February 1971 that the Provos started an aggressive campaign and ambushed British Army patrols. The British Army's robust screen-and-search tactics and internment policy did a great recruiting job for the Provos, who consolidated their power in the no-go areas so that by 1972 their active strength was about 1,500.

Seeing Catholic support shifting to John Hume, the Social Democratic Labour Party, and the hope of a political solution, MacStiofain instituted a bombing campaign in 1972 aimed at members of the British Army, the UDR and the RUC. The objective was to rile the British Army into reprisals, alienating Catholics further from politics and drawing them to support the Provos. The car bomb campaign, aimed at economic and political targets and central Belfast, had the effect of tying troops down on guard duty, which reduced the British Army's counterinsurgency patrolling. In addition, land mines and bombs were distributed widely through Northern Ireland to give the impression of omnipresence.

Having caught wind of Operation Motorman, which was designed to mop them up in the no-go areas, the Provo command determined that they were not equipped for a static defense against armored force and that such a defense was "contrary to the principles of guerrilla warfare and IRA experience" (MacStiofain 1975, 298). They believed they should avoid fighting in a confined area against superior numbers and equipment. And so the IRA melted away on July 3, 1972, as the British Army advanced into the Bogside, the Falls, Portadown, Lurgan and Armagh. Only two people were shot in the whole operation. Following the dictum of Sun Tzu—"Uproar in the east, strike in the west"—with the British Army tied up in urban no-go areas, the Provos had a field day in the countryside. They adopted a "see-saw" strategy. Inducing a lapse of readiness in the towns by striking in the countryside, they would then strike in the no-go areas, supposedly cleared by Motorman.

However, by 1977, the British Army had the upper hand in Belfast with helicopters tied in with patrols, stakeouts and undercover operations. The Provos maintained the initiative in the rural periphery. In the face of these circumstances, the IRA began to reorganize for "the long way," the Irish equivalent of *guerra prolongada*. The old strategy of "one more heave" to throw off the British yoke was obviously unrealistic. The old structure of battalions and brigades was abandoned in favor of a cellular structure. The reduced active strength of 300 was organized into four-person, active service units (ASU), directed by nine operational officers, dedicated to surviving in a long campaign of attrition. The ASUs had about 3,000 supporters, who provided safe houses, lookouts and transport. This cellular structure was less dependent on popular support and, thus, was less vulnerable to penetration by informers. Their independence also cut down on their sympathetic ties with the Catholic community. Conscious of this loss of passive support in 1981, the new leadership announced the double strategy of seeking victory with the ballot paper in one hand and an Armalite in the other. Gerry Adams' election as the Sinn Fein Member of Parliament in 1983 signalled the first success in this strategy.

The gunmen continue to be deployed mostly in the rural border regions of Fermanagh, Tyrone and Armagh, where they attack the British Army and the security system. The Glover Report clearly believed that the IRA had the wherewithal to continue its military operation into the foreseeable future with what are dedicated professional soldiers.

TACTICS

British Tactics

In its early confrontations with large numbers of Catholics expressing their contempt for civil and military authority, the British Army could not use the riot tactics that had been used in India, Hong Kong, Singapore and Aden. The selective shooting of the more active members of the front rank of the mob, the young Derry hooligans (YDHs) of the Bogside in British Army parlance, would not go down well before the cameras, nor would it help sustain the myth of common law civility. Every soldier was issued with a Yellow Card which specified when "reasonable force" could be used in the prevention of a crime in compliance with section 3 of the Criminal Law Act of 1967. The devices of CS gas and rubber and plastic bullets were employed liberally, and the tactic of the "snatch squad" was devised to grab the most prominent rioters.

The shooting war with the IRA was effectively provoked by cordon-and-sweep operations and destructive house searches for weapons in July 1970. This plus internment in October 1970 really got the IRA off the ground. Having generated dissidence, the British Army then turned Kitson's counterinsurgency tactics against the IRA. These tactics are based on the strategic premises that the British Army could no longer rely on force of numbers in the face of possible revolution and that they needed a greater facility for psychological and irregular warfare and the creation and manipulation of information. Aside from disseminating false information and manipulating the media and the social services, military operations were geared to infiltration and systematic surveillance. Patrol tactics, community projects and interrogations were all geared to gathering low-level intelligence: names and places. Company commanders were admonished to know their "patches." A fluid infiltration of the host community of the insurgents and plainclothes operatives were in action by 1972, partly replacing the conventional ring of steel tactics of Tuzo. Armed plainclothes patrols—called military reconnaissance forces and trained by the SAS—undercover units, *agents provocateurs* and false front operations, like the Four Square Laundry, helped Kitson "make war with a filing cabinet." By 1973 the British Army had

500,000 intelligence files, which is about the same number as the Catholic population. They had an index card on every Catholic boy over the age of twelve and every girl over fourteen. For south Armagh, there was a dossier on every inhabitant of Crossmaglen, whether they were suspected of IRA involvement or not. By 1977 all of this was computerized at British Army headquarters at Lisburn.

By 1973 patrol tactics in hostile areas had evolved from one column of 300 men in a cloud of CS gas, through five platoons moving along five parallel streets, to five "bricks" of five men apiece, moving through an area in radio contact, seeking to maximize Provo uncertainty. Soldiers learned when ambushed not to take cover to return fire but to enter houses quickly to seek snipers. In the rural scene, the SAS were brought in officially in 1976 as a spearhead battalion, operating in small, highly skilled units, less vulnerable to land mines and rocket launchers. In areas where they achieved the upper hand, in Belfast by 1978, the British Army used air-land tactics with constant helicopter surveillance in touch with plainclothes units on the street and riflemen staked out throughout the city suburbs. IRA members were apprehended trying to buy Stingers in Miami, Florida, in 1990.

Provo Tactics

After the initial phase of spontaneous, barricades-manning, community defense and a bombing campaign aimed at local economic targets, from early 1971, the Provos began to fight a protracted war with government forces in the cities and countryside. Internment provided not only the incentive for joining the IRA, but also the school for the development of urban guerrilla tactics. What emerged were local adaptations of the usual recourse of insurgents in the face of a superior force—ambush, sniping, assassination and terrorist bombing. This is the offensive defense, which employs the terrain and the population for cover and concealment and refuses to be drawn into a pitched battle where superior firepower would tell. Terror was a major weapon used to create fear and uncertainty. Bombing increased the cost of government and drove a wedge between the Catholic population and the political process. Fearful and irritated reaction by government forces maintained Catholic alienation. This destructive irritation was extended farther afield in the autumn of 1976 when the bombing campaign was carried to Great Britain. In February 1991 the IRA actually made a mortar attack on 10 Downing Street and on the British War Cabinet.

In August 1972 MacStiofain introduced the one-shot sniper tactic in Derry, Belfast and south Armagh, exploiting densely textured terrain and a sympathetic population to the utmost. Local boys were used to scout out an ambush site. The gunman would move into position and wait for someone to bring him a weapon. On sighting a British patrol he would fire one shot, leaving the scene and the weapon for someone else to pick up. The favorite weapon was the Armalite, the civilian version of the American MI6 assault rifle, a high-velocity semiautomatic, which is small and easily concealed.

For the long haul of continuing, low-level conflict since the late 1970s, the war has been fought between two sets of professionals. The reorganization of the IRA in the latter part of 1978 into ASUs brought a distinct change of tactics. The cells began to operate independently of the local population, mostly, for security's sake, outside the home areas of its members. This diminished the potential spontaneity of their attacks. The IRA turned to once-off, highly planned bomb attacks like Mullaghmore and Warrenpoint in August 1979 and Camlough in May 1981. The bombing campaign was extended in the late 1980s from Great Britain to British military installations, personnel and their families in Germany and Belgium. Gibralter seems to have been another target. The close involvement of fighters with the people has been severed and replaced by the political tactics of Sinn Fein, leaving "a stratum of intelligent, astute and experienced terrorists," to quote the Glover Report, at war with the British Army and police. In November 1989 Peter Brooke, the secretary of state for Northern Ireland, attracted a storm of protest from Unionists and loyalists when he admitted that military defeat of the Provos was not in sight and that "a policy of containment" was the only prospect. He hinted at willingness to talk to Sinn Fein: "If in fact the terrorists were to decide that the moment had come when they wished to withdraw from their activities" (Hearst and Wintour, 1989).

CONCLUSION

Both the evidence of the geography of violent events and the strategic statements of British authorities and the IRA bear a strong resemblance to the prescriptions of the gradient of power model. The incumbents hold the center and contain the insurgents on the periphery. The insurgents survive most easily in the outer fringes but must press into the core area when they can to erode and question the government's ability to govern. This pattern, established by 1974, has survived for more than a decade. The low level of this equilibrium of violence, giving Northern Ireland a lower death rate than most major U.S. cities, is accompanied by continued competition by more politic means as well as a terrorist campaign that extends to Great Britain and the European mainland. In the turn to politics and the prospect of negotiation, as well as the demise of Northern Ireland's strategic significance, lies the hope for a solution.

REFERENCES

Baynard, P. 1985. "The Security Forces 1969–1985." In *The Divided Province: The Troubles in Northern Ireland 1969–1985*, edited by K. Jeffery. London: Orbis.

Bew, P., and H. Patterson. 1985. *The British State and the Ulster Crisis: From Wilson to Thatcher*. London: Verso.

Boal, F., and J. Douglas (eds). *Integration and Division: Geographical Perspectives on the Northern Ireland Problem*. Particularly P. Robinson, "Plantation and Colonization: The Historical Background"; R. Buchanan, "The Planter and the Gael: Cultural Dimensions of the Northern Ireland Problem"; P. Compton, "The Demographic Dimensions of Integration and Division of Northern Ireland"; J. Douglas, "Northern Ireland: Spatial Frameworks and Community Relations"; R. Murray, "Political Violence in Northern Ireland 1969–77." London: Academic Press.

Davis, R. 1986. "Kitson versus Marighela: The Debate over Northern Ireland Terrorism." In *Ireland's Terrorist Dilemma*, edited by Y. Alexander and A. O'Day. Dordrecht, the Netherlands: Martinus Nijhoff.

Downey, J. 1983. *Them and Us: Britain, Ireland and the Northern Question 1969–1982*. Dublin: Ward River Press.

Hamill, D. 1985. *Pig in the Middle: The Army in Northern Ireland 1969–1984*. London: Methuen.

Hearst, D., and P. Wintour. 1989. "Fury at 'Talks with IRA' Hint by Brooke." *Manchester Guardian Weekly*, November 12.

Kitson, F. 1971. *Low Intensity Operations: Subversion, Insurgency, Peace-Keeping*. London: Faber, p. 24.

MacStiofain, S. 1975. *Revolutionary in Ireland*. London: Gordon Cremonesi.

12

FIGHTING IN THE LANDSCAPE AND FIGHTING FOR A PLACE

The motives and behavior of those who start wars were examined in Chapters 4 and 5. It is evident that simplified and mythical versions of their attitudes infect the officer class with enthusiasm for war. Military education includes an indoctrination in patriotism, class spirit and identity with the military caste. Unless there is a strong connection between the objects and passions of the ruling elite and its warrior chiefs, the political structure is insecure. This leaves the talents and goals of the common soldiery to be discussed. It seems appropriate to end a discourse on geography and warfare by examining the aptitudes and attitudes of those who bear the main brunt of the fighting, killing and dying. Some brief reference has been made to this in the context of guerrilla tactics, but this needs extension and generalization to all varieties of combatant. What is it that induces people to undergo great hardship, risk injury and death and kill other people—often in a setting far from home and any personal sense of defending turf—and what makes some better at it than others?

In the animal world, the needs of genetic survival dictate individual urges to fight for territorial and copulatory rights. When we consider the organized fighting of mankind, the spoils are far less tangible nowadays. There is in many of us a carryover of joy in the physical and mental exertions of battle and in some of us a delight in killing. In general, however, we are repelled by the prospect of killing other people. This repulsion is overcome, and the pleasure in strenuous competition is harnessed, collectivized and reinforced by fear, training and the arrangements and ethos of soldiering. This has mostly been divorced from specific territoriality; however, vestiges of this are retained for example, in British Army regimental identities. An intimate relation with the ground for which one fights has been subsumed in the abstraction of patriotism or professionalism.

William McNeill identified an underlying ambivalence in the present attitude to war. Through history the collective spirit has seen its highest expression as residing in warrior heroism, self-sacrifice, bonding and prowess. There is an increasing incongruity between this tradition and the growing repugnance for military killing and destruction. This has been heightened in the last half century by the quantum leap in the available power to destroy, maim and kill. The horrors of war are greater because it involves killing more and more women and children. In mid-century half those killed in war were civilians. Now civilians account for three-quarters of the death toll of war.

John Keegan pointed out that those in the war machine, who can exert a choice in what they do, have progressively removed themselves from the business of butchery since the arrival of the gun on the battlefield. Apart from the remote romance of war in the air, the officer class has diminished the prospects of having to engage in the act of killing anyone. They have reduced their weapons to symbolic toys and are content to direct the action. Whether this evolution was a matter of choice or was forced upon them by the increasing complexity of command and control requiring their full attention is arguable, but the effect is the same.

This leaves us with the talents of the common soldier on the ground, and what moves him to kill, be he a regular or a partisan. In this we can recognize two aspects to his involvement with the geographical dimensions of the human condition. On the one hand, there are the skills required for coping with distance and geographical reality. Beyond this range of spatial talent, there is the domain of emotional involvement. There is a degree to which soldiers' lives are governed by their sentiment for and identity with geographical entities.

SPATIAL ABILITY

Coping with the spatial dimension of war takes places over a number of different ranges of distance at which the various aspects of violent conflict operate. Each range demands different skills and responses. There are distinct scales of spatial activity and ability ranging up from the mental powers of geometric abstraction, through graphic ability, the skills of bodily coordination and the power of visual perception, to the geographic talent for conjuring up an image of what lies beyond the range of unaided vision in the battlefield. The first and last of these scales of operation call for the exercise of mental powers of abstraction and manipulation, which are neither everyday nor universally developed. Whether the sense of distance and location is hardwired into our brain or not is unclear. It does seem, however, that locational sense derives very readily for innate spatial skills. The talent for geographic comprehension and representation develops with very little instruction and encouragement. Children will draw recognizable maps of their environs with very little prompting. There is obviously much about our physical and mental makeup that is geared to increasing our range of vision and operations because we were slow and weak hunters and gatherers. The most materially primitive societies have very rich geographic doctrines and lore. The "song lines" of Australian aborigines are an obvious case in point. There is a realm of the mind in which we construct and store images of the land. The next step is when we realize these on sand, paper, blackboard or cathode ray screen.

There is some evidence of a division of people into those inclined to pictures—the spatial and intuitive—and those inclined to symbols—the verbal and logical. Whether mental and manual spatial skill carries over into the domain of bodily movement and the coordination of eye and hand is not fully explored. There is a possible genetic connection between left-handedness, maleness, proneness to dyslexia and spatial rather than verbal aptitudes. It seems obvious that a soldier needs to be skillful spatially in order to survive. Good vision and positional sense are great assets. These qualities involve not only the physical advantage of good peripheral vision, but also the ability to read positions and velocities of others in the landscape and to project the development of the configuration over an appropriate time for decision. Some people have a better feel for the angles and can see interception or avoidance paths intuitively. The perceptual skills of estimating distances and evaluating features in the landscape over a range of a mile or more vary widely among the population. There is much lore to be learned on finding your way without a map or a compass. A church steeple can be made out up to nine miles away. At half a mile, a man looks like a post; at fifty yards, the whites of his eyes can be seen. Directional cues need a local feel for nature in the countryside. Which side of the tree is mossy depends on the prevailing direction of moisture-bearing winds. Snow ridges and sand dunes give similar clues in Arctic and desert settings. In the townscape the direction of television aerials and the patterns of rusting, paint peeling and pavement wear can help one find the way. Smell as well as sight comes into play. The desert geomorphologist who founded the Long Range Desert Group, Brigadier R. A. Bagnold, found water in rough country because he picked up the smell of a camel eight miles away.

Beyond the horizon there is a switch from perceptual to cognitive skills, and we enter the domain of the geographer and the synthetic talent which von Clausewitz called *Ortsinn*, a sense of place (Greene 1943). This involves "the power of quickly forming a correct geometrical idea of any portion of the country and consequently of being able to find one's place in it at any time." It implies the ability to summon up an accurate mental map of the landscape and to employ it in making tactical and strategic decisions. Beyond the range of actual fighting, much in the way of strategic decision is amenable to calculation. The time frame is not the split instant of combat, but it allows analytic formulation and solution. So logistic problems in the shorter run, and the design of geographical systems of weapon deployment in the longer run, can be submitted to the rigors of mathematics to find optimal solutions.

AFFECTION, IDENTITY AND TERRITORY

It could be said that the persistence of war as a means of conflict resolution reflects the fact that we still carry the emotional baggage of hunters and gatherers. A mass of town dwellers get dragged into violent action in competition for turf, even though most of us are not only far removed from the hunting grounds but also from the fields. Among rural insurrectionists there may be a strong and real sense of fighting for the acres that give one life. The sense of *patria chica* may provide the incentive for savage bravery. Possibly an urbanized form of this is at work in Beirut, in the townships of South Africa and among the street gangs of New York and Los Angeles.

Most soldiers are completely divorced from their home turf in action. Indeed, this created a significant health problem in the early stages of modern warfare. The term *nostalgia* was coined for the physical homesickness of German peasants dragged off to play soldier in foreign fields in the seventeenth century. What arose to counter the natural reluctance to fight for someone else's fields was the inculcation of a professional ethos for soldiers and the cultivation of a regard for the honor of the unit.

The increasing sense of popular political involvement of the last two centuries has, however, generalized and abstracted the affection for place and geographical identity and has created in nationalism the most powerful sentiment for violence. This is the emotional force directed to territory that has overwhelmed both the peasants' love of their native soil and the elite appreciation of the charm of bits of landscape. Over the last 200 years all other attachments and sentiments have been subordinated to the love of country. This is the geographical identity that overrides love of God and mankind on the one hand, and love of home, family and friends on the other. This is the sentiment that led millions to numb their minds and bodies to instinct and familiar ties and to sacrifice life and limb for a geographical abstraction.

The division of the world into *us* and *them*, objects of love on the one hand and, at least, indifference on the other, has usually involved geographical boundaries. The radius of affection has been extended from the family to the tribe, to the city-state and, finally, to the nation. This is the identity with place for which people have suppressed feelings of home, neighborhood and region on the one hand, and class, religion or humanity on the other. Nationality as it took shape following the French Revolution, transferring political identity from a geographical entity to the mass of people in the territory, is clearly the mightiest collective emotion we can observe at work in the world today. Although we may have vestigial feelings for the place where we were born or where our parents live, the words of anthems, pledges, lessons, speeches and advertisements have worked toward the identification of "home" with "nation." With migration, residential restlessness, greater mobility and greater, at least vicarious, experience of the rest of the world on television, nostalgia is no longer a serious malady, merely a sloppy sentiment.

What people suspend their instincts and their intellectual powers for and, indeed, are willing to put themselves in harm's way for, is the nation. This is a geographical entity and a population that people can know personally only in the abstract. This abstraction has been obscenely perverted by those who seek power over others. Millions have responded to the exhortation to kill "them," whose humanity is denied, and to be killed for the greater glory of "us." The countervailing religious or scientific identity with all humanity, or the political identity with your economic peers, fall by the wayside for most when the passions are aroused and the most atrocious inhumanity finds its excuse.

Apart from dire economic necessity or the compulsion of the law, it is the geographical abstraction embodied in patriotism that gets large numbers of people into a position to fight and kill. However, when it comes to the moment of truth, the urge to combat seems to arise from a very local sense of community and necessity on the battlefield. Although there is now a raging controversy over S.L.A. Marshall's assertion that less than a quarter of infantrymen will fire their weapons at the enemy, it is evident that the needs of survival outweigh the will to victory (Spiller 1988). For the soldier, survival in a nasty and dangerous setting focuses his attention on memberships of a group of half a dozen comrades in a very small tract of the landscape. Keegan points out that the topography of the battlefield plays a major role in motivating soldiers to fight and kill. Pushed into combat by the threat of punishment if not death, or carried along on a wave of finer emotion, people find themselves in deadly contest with others over a piece of ground that offers some protection and an avenue for retreat or reinforcement. For this, people will learn to kill skillfully. War, then, becomes the contrivance and manipulation of a large number of such contests in such a manner that the advantage will reside with your forces. With a reduction in the compulsory power of the state or the emotional force of patriotism, there is a prospect that fewer people will find themselves pushed into such extremities of feeling and action. Perhaps sometime the nation and the landscape will become objects of love and affection and no longer the goals and setting for murderous competition.

REFERENCES

Greene, J. I. 1943. *The Living Thoughts of Clausewitz.* New York: Longman, Greene and Co., p. 25.

Keegan, J. 1976. *The Face of Battle*. London: Penguin.
Kohn, H. 1961. *The Idea of Nationalism*. New York: Macmillan Co.
McNeill, W. H. 1982. *The Pursuit of Power: Technology, Armed Force, and Society Since AD 1000*. Chicago: University of Chicago Press.
Spiller, R. J. 1988. "S.L.A. Marshall and the Ratio of Fire". *Royal United Services Institute Journal* (Winter): 63–71.

BIBLIOGRAPHY

Addington, L. H. 1984. *The Patterns of War Since the Eighteenth Century*. Bloomington: Indiana University Press.

Alger, J. I. 1982. *The Quest for Victory: The History of the Principles of War*. Westport, Conn.: Greenwood.

———. 1985. *Definitions and Doctrine of the Military Art Past and Present*. Wayne, N.J.: Avery.

Anthony, P. 1989. *Pilots and Rebels: The Use of Aircraft in Unconventional Warfare, 1919–1988*. London: Brassey's.

Asprey, R. B. 1975. *War in the Shadows: The Guerrilla in History*. 2 vols. New York: Doubleday.

Azan, P. 1918. *The Warfare of Today*. Boston: Houghton Mifflin.

Bacon, R., J.F.C. Fuller, and P. Playfair, eds. 1944. *Warfare Today*. London: Odhams.

Bailey, J.B.A. 1989. *Field Artillery and Firepower*. Oxford: The Military Press.

Barnett, C. 1986. *Old Battles and New Defences: Can We Learn from Military History?* London: Brassey's.

Baynard, P. 1985. "The Security Forces 1969–1985". In *The Divided Province: The Troubles in Northern Ireland 1969–1985*, edited by K. Jeffery. London: Orbis.

Beckett, J. F. 1988. *The Roots of Counter Insurgency: Armies and Guerrilla Warfare, 1900–1945*. London: Blandford.

Bernhardi, F. von. 1920. *The War of the Future in Light of the Lessons of the World War*. London: Hutchinson.

Betnor, R. 1969. *Decisive Warfare: A Study in Military Theory*. Harrisburg, Pa.: Stackpole.

Bidwell, S., and D. Graham. 1982. *Firepower: British Army Weapons and Theories of War, 1904–1945*. London: George Allen and Unwin.

Blaufarb, D. 1977. *The Counterinsurgency Era: U.S. Doctrine and Performance 1950 to the Present*. New York: The Free Press.

Blaufarb, D., and G. Tanham. 1989. *Who Will Win? A Key to the Puzzle of Revolutionary War*. New York: Crane Russak.

Boulding, K. 1963. *Conflict and Defense*. New York: Harper and Row.

Bucholz, A. 1985. *Hans Delbruck and the German Military Establishment*. Iowa City: University of Iowa Press.

Carver, M. 1976. *The War Lords: Military Commanders of the Twentieth Century*. London: Weidenfield and Nicolson.

———. 1980. *War Since 1945*. London: Weidenfeld and Nicolson.

Chaliand, G. 1987. *Terrorism: From Popular Struggle to Media Spectacle*. London: Sayi.

Chandler, D. 1987. *The Dictionary of Battles*. London: Ebury Press.

Chapman, J. 1989. *French Power in Africa*. Oxford: Basil Blackwell.

Citano, R. M. 1987. *The Evolution of Blitzkrieg Tactics: Germany Defends Itself against Poland, 1918–1933*. Westport, Conn.: Greenwood.

Cordesman, A. H., and A. Wagner. 1990. *The Lessons of Modern War*. 3 vols. Boulder, Colo.: Westview.

Creasy, E. S. 1852. *Fifteen Decisive Battles of the World: From Marathon to Waterloo*. Reprint 1987. New York: Military Heritage Press.

Davidson, B. 1981. *The People's Cause: A History of Guerrillas in Africa*. London: Longman.

Davis, R. 1986. "Kitson versus Marighela: The Debate over Northern Ireland Terrorism". In *Ireland's Terrorist Dilemma*, edited by Y. Alexander and A. O'Day. Dordrecht, the Netherlands: Martinus Nijhoff.

Debray, R. 1967. *Revolution in the Revolution? Armed Struggle and Political Struggle in Latin America*. New York: Monthly Review Press.

Dewar, M. 1984. *Brush Fire Wars: Minor Campaigns of the British Army Since 1945*. New York: St. Martin's.

Drea, E. J. 1981. *Nomohan: Japanese-Soviet Tactical Combat, 1939*. Ft. Leavenworth, Kans.: U.S. Army Command and General Staff College.

———. 1984. *Defending the Driniumor: Covering Force Operations in New Guinea, 1944*. Ft. Leavenworth, Kans.: U.S. Army Command and General Staff College.

Duffy, C. 1987. *The Military Experience in the Age of Reason*. London: Routledge and Keegan Paul.

du Picq, C. A. 1978. *Etudes sur le combat*. Paris: Editions Champ Libre.

Elliot-Bateman, M. 1967. *Defeat in the East: The Mark of Mao Tse-Tung on War*. London: Oxford University Press.

English, J. A. 1981. *On Infantry*. New York: Praeger.

Evans, E. 1987. *Wars without Splendor: The US Military and Low Level Conflict*. Westport, Conn.: Greenwood.

Farrar-Hockley, A. 1988. *Opening Rounds: Lessons of Military History, 1918–1988*. London: Deutsch.

Fuller, J.F.C. 1925. *The Foundation of the Science of War*. London: Hutchinson.

Gabel, C. R. 1985. *Seek and Destroy: US Army Tank Destroyer Doctrine in World War II*. Ft. Leavenworth, Kans.: U.S. Army Command and General Staff College.

Galvin, J. R. 1969. *Air Assault: The Development of Airmobile Warfare*. New York: Hawthorn.

Galvla, D. 1964. *Counter-Insurgency Warfare: Theory and Practice*. New York: Praeger.

Gat, A. 1989. *The Origins of Military Thought: From Enlightenment to Clausewitz.* Oxford, England: Clarendon.

Glantz, D. M. 1983. *August Storm: The Soviet 1945 Strategic Offensive in Manchuria.* Ft. Leavenworth, Kans.: U.S. Army Command and General Staff College.

———. 1984. *The Soviet Airborne Experience.* Ft. Leavenworth, Kans.: U.S. Army Command and General Staff College.

Glover, M. 1980. *Warfare from Waterloo to Mons.* London: Cassell.

———. 1980. *Warfare in the Age of Bonaparte.* London: Cassell.

———. 1982. *The Velvet Glove: The Decline and Fall of Moderation in War.* London: Hodder and Stoughton.

Greene, J. I. 1943. *The Living Thoughts of Clausewitz.* New York: Longman, Greene.

Greene, T. N., ed. 1962. *The Guerrilla and How to Fight Him.* New York: Praeger.

Gregory, B. 1989. *Mountain and Arctic Warfare.* London: Patrick Stephens.

Griffith, P. 1981. *Forward into Battle: Fighting Tactics from Waterloo to Vietnam.* Sussex, England: Anthony Blond.

———. 1989. *Military Thought in the French Army, 1815–51.* Manchester, England: Manchester University Press.

Guevara, C. 1985. *Guerrilla Warfare.* Lincoln: University of Nebraska Press.

Hamill, D. 1985. *Pig in the Middle: The Army in Northern Ireland 1969–1984.* London: Methuen.

Harkavy, R., and S. G. Neuman, eds. 1985. *The Lessons of Recent Wars in the Third World.* 2 vols. Lexington, Mass.: D. C. Heath.

Hartcup, G. 1979. *Camouflage: A History of Concealment and Deception in War.* London: David and Charles.

Heilbrunn, O. 1962. *Partisan Warfare.* New York: Praeger.

Heller, C. E. 1984. *Chemical Warfare in World War I: The American Experience, 1917–1918.* Ft. Leavenworth, Kans.: U.S. Army Command and General Staff College.

Horton, E. R., and K. Vunson, eds. 1989. *Jane's Battlefield Surveillance, 1988–90.* London: Jane's Defence Data.

House, J. M. 1984. *Towards Combined Arms Warfare: A Survey of Twentieth Century Tactics, Doctrine and Organization.* Ft. Leavenworth, Kans.: U.S. Army Command and General Staff College, Combat Studies Institute. Research Survey No. 2.

Isby, D. 1981. *Weapons and Tactics of the Soviet Army.* London: James.

Jacobson, C. G., ed. 1987. *The Uncertain Course: New Weapons, Strategies and Mind Sets.* London: Oxford University Press.

Joes, A. J. 1986. *From the Barrel of a Gun: Armies and Revolutions.* London: Pergamon-Brassey's.

Jones, D. R. 1984. *The Advanced Guard and Mobility in Russian and Soviet Military Thought and Practice.* Gulf Breeze, Fla.: Academic International Press.

Keegan, J. 1976. *The Face of Battle.* London: Penguin.

Keegan, J., and A. Wheatcroft. 1986. *Zones of Conflict: An Atlas of Future Wars.* New York: Simon and Schuster.

Kiernan, V. G. 1985. *European Empires from Conquest to Collapse, 1815–1960.* Leicester, England: Leicester University Press.

King, M. J. 1985. *Rangers: Selected Combat Operations in World War II*. Ft. Leavenworth, Kans.: U.S. Army Command and General Staff College.

Kitson, F. 1971. *Low Intensity Operations: Subversion, Insurgency, Peace-Keeping*. London: Faber.

Laffin, J. 1989. *The World in Conflict 1989: War Annual 3: Contemporary War Described and Analyzed*. London: Brassey's.

Laqueur, W. 1976. *Guerrilla: A Historical and Critical Study*. Boston: Little, Brown & Co.

Larson, R. J. 1984. *The British Army and the Theory of Armored Warfare, 1918–1940*. Newark: University of Delaware Press.

Lewis, S. J. 1985. *Forgotten Legions: German Army Infantry Policy, 1918–1941*. New York: Praeger.

Liddell Hart, B. 1967. *Strategy*. New York: Praeger.

Lupfer, T. T. 1981. *The Dynamics of Doctrine: The Changes in German Tactical Doctrine During the First World War*. Ft. Leavenworth, Kans.: U.S. Army Command and General Staff College.

Luttwak, E. N. 1985. *Strategy and History*. 2 vols. New Brunswick, N.J.: Transaction Books.

McColl, R. 1969. "The Insurgent State: Territorial Bases of Revolution." *Annals of the Association of American Geographers* 59(4).

McCuen, J. S. 1966. *The Art of Counter-Revolutionary War: The Strategy of Counter-Insurgency*. Harrisburg, Pa.: Stackpole.

McElwee, W. 1974. *The Art of War: Waterloo to Mons*. Bloomington: Indiana University Press.

McInnes, C., and G. D. Sheffield. 1988. *Warfare in the Twentieth Century: Theory and Practice*. London: Unwin, Hyman.

McNeill, W. H. 1982. *The Pursuit of Power: Technology, Armed Force, and Society Since AD 1000*. Chicago: University of Chicago Press.

MacStiofain, S. 1975. *Revolutionary in Ireland*. London: Gordon Cremonesi.

Mallin, J., ed. 1970. *Strategy for Conquest: Communist Documents on Guerrilla Warfare*. Coral Gables, Fla.: University of Miami Press.

Mao Tse Tung. 1961. *On Guerrilla Warfare*, translated by S. B. Griffith. New York: Praeger.

Marshall, S.L.A. 1947. *Men Against Fire*. New York: Morrow.

Martic, M. 1975. *Insurrection: Five Schools of Revolutionary Thought*. New York: Donnellan.

Millet, A. R., and W. Murray, eds. 1987. *Military Effectiveness*. 3 vols. Boston: George Allen and Unwin.

Montross, L. 1960. *War through the Ages*. New York: Harper and Row.

Osanka, F. M. 1962. *Modern Guerrilla Warfare: Fighting Communist Guerrilla Movements, 1941–1961*. New York: The Free Press.

O'Sullivan, P. 1986. *Geopolitics*. London: Croom Helm.

O'Sullivan, P., and J. Miller. 1983. *The Geography of Warfare*. London: Croom Helm.

Paget, J. 1967. *Counter-Insurgency Operations: Techniques of Guerrilla Warfare*. New York: Walker.

Paret, P., and J. W. Shy. 1962. *Guerrillas in the 1960s*. New York: Praeger.

Parker, G. 1988. *The Military Revolution: Military Innovation and the Rise of the West*. Cambridge: Cambridge University Press.

Perret, B. 1989. *Canopy of War: Jungle Warfare from the Earliest Days of Forest Fighting to the Battlefields of Vietnam*. London: Patrick Stephens.

Peltier, L. C., and G. E. Pearcy. 1966. *Military Geography*. Princeton, N.J.: Van Nostrand.

Pomeroy, W. J., ed. 1968. *Guerrilla Warfare and Marxism*. New York: International Publishers.

Pike, D. 1986. *PAVN: People's Army of Vietnam*. Novato, Calif.: Presidio.

Pustay, J. S. 1965. *Counterinsurgency Warfare*. New York: The Free Press.

Quick, J. and P. O'Sullivan. 1988. "Military Analysis of Urban Terrain." *Professional Geographer* 38(3): 286–90.

Regan, G. 1987. *Someone Had Blundered . . . A Historical Survey of Military Incompetency*. London: B. T. Batsford.

Resendes, S., and P. O'Sullivan. 1988. "Terrain and the Effectiveness of Guerrillas." *Papers and Proceedings of Applied Geography Conferences* 9:234–239.

Rice, E. 1988. *Wars of the Third Kind: Conflict in Underdeveloped Countries*. Berkeley: University of California Press.

Robertson, W. G. 1965. *Counterattack on the Naktong, 1950*. Ft. Leavenworth, Kans.: U.S. Army Command and General Staff College.

Rothenburg, G. 1978. *The Art of Warfare in the Age of Napoleon*. Bloomington: Indiana University Press.

Sarkesian, S. C. 1980. *Nonnuclear Conflicts in the Nuclear Age*. New York: Praeger.

———. 1984. *America's Forgotten Wars: The Counter Revolutionary Past and Lessons for the Future*. Westport, Conn.: Greenwood.

———. 1986. *The New Battlefield: The United States and Unconventional Conflicts*. Westport, Conn.: Greenwood.

Sarkesian, S. C., and W. L. Scully, eds. 1981. *US Policy and Low Intensity Conflict: Potentials for Military Struggles in the 1980s*. New Brunswick, N.J.: Transaction Books.

Shackley, T. 1981. *The Third Option: An American View of Counterinsurgency Operations*. New York: McGraw Hill.

Shaeffer, R. K., ed. 1989. *War in the World System*. Westport, Conn.: Greenwood.

Shrader, C. 1982. *Amicide: The Problem of Friendly Fire in Modern War*. Ft. Leavenworth, Kans.: U.S. Army Command and General Staff College.

Shultz, R., R. Pfaltzgraff, U. Ra'anan, W. J. Olson, and I. Lukes, eds. 1989. *Guerrilla Warfare and Counterinsurgency: US-Soviet Policy in the Third World*. Lexington, Mass.: D. C. Heath.

Sivard, R. 1989. *World Military and Social Expenditures*. Washington, D.C.: World Priorities Inc.

Smith, E. D. 1988. *Victory of a Sort*. London: Robert Hale.

Spiller, R. J. 1981. *"Not War but Like War": The American Intervention in Lebanon, 1958*. Ft. Leavenworth, Kans.: U.S. Army Command and General Staff College.

———. 1988. "S.L.A. Marshall and the Ratio of Fire." *Royal United Services Institute Journal* (Winter): 63–71.

Stelmach, D. S. 1978. *The Influence of Russian Armored Tactics on the North Korean Invasion of 1950*. Ann Arbor, Mich.: University Microfilms.

Strachan, H. 1983. *European Armies and the Conduct of War*. London: George Allen and Unwin.

Taylor, S. 1986. *Social Science and Revolutions*. New York: St. Martin's.

Thompson, L. B., ed. 1989. *Low Intensity Conflict: The Pattern of Warfare in the Modern World*. Lexington, Ky.: D. C. Heath.

Thompson, R. 1966. *Defeating Communist Insurgency: The Lessons of Malaya and Vietnam*. New York: Praeger.

Thompson, R., ed. 1981. *War in Peace: An Analysis of Warfare Since 1945*. London: Orbis.

Townshend, C. 1986. *Britain's Civil Wars: Counterinsurgency in the Twentieth Century*. London: Faber and Faber.

Travers, R., and C. Archer, eds. 1982. *Men at War: Politics, Technology and Innovation in the Twentieth Century*. Chicago: Precedent.

U.S. Department of the Army. 1979. *Field Manual 90–10: Military Operations in Urbanized Terrain (MOUT)*. Washington, D.C.: U.S. Government Printing Office.

U.S. Department of the Army. 1982. *Field Manual 100–5: Operations*. Washington, D.C.: U.S. Government Printing Office.

Wade, G. H. 1984. *Rapid Deployment Logistics: Lebanon, 1958*. Ft. Leavenworth, Kans.: U.S. Army Command and General Staff College.

Wallach, J. H. 1986. *The Dogma of the Battle of Annihilation: The Theories of Clausewitz and Schlieffen and their Impact on the German Conduct of Two World Wars*. Westport, Conn.: Greenwood.

Weigley, R., ed. 1975. *Fire and Movement: Bargain-Basement Warfare in the Far East*. New York: Thomas Y. Crongil.

Wheatcroft, A. 1983. *The World Atlas of Revolutions*. New York: Simon and Schuster.

White, C. E. 1989. *The Enlightened Soldier: Scharnhost and the Militarische Gesellschaft in Berlin*. New York: Praeger.

Winton, H. R. 1988. *To Change an Army: General Sir John Burnett Stuart and British Armored Doctrine, 1917–1938*. Lawrence, Kans.: University of Kansas Press.

Wray, T. A. 1984. *Standing Fast: German Defensive Doctrine on the Russian Front During World War II*. Ft. Leavenworth, Kans.: U.S. Army Command and General Staff College.

Yale, W., I. D. White, and H. von Manteuffel. 1970. *Alternative to Armageddon: The Peace Potential of Lightning War*. New Brunswick, N.J.: Rutgers University Press.

INDEX

Abadan, 11, 13, 27
Acheson, Dean, 55
active service units (ASU), 158, 161
Addis Ababa, 6, 13, 44, 54
Afghanistan, 5, 7, 15, 26, 29, 44, 51, 58, 59, 115
Africa, 35, 37
African National Congress (ANC), 52, 51
Agincourt, 90
Ahmar Mts., 13
airborne insertion, 81, 99
air-land tactics, 161
Alexander, 89, 110
Algeria, 6, 8, 11, 41, 44, 54, 55
Algiers, 97
Allenby, Edmund, 114
Alps, 27, 32
Alsace Lorraine, 95, 96
Amazon, 34
American Civil War, 9, 94
American Revolution, 73, 78
amphibious operations, 98, 99
Amu Darya River, 7, 15
Andes, 67
Angola, 44, 52, 53
Anti-Lebanon Mts., 15
Antiochus, 89
apartheid, 51
Aquino, Corazon, 62, 63, 148
Arbela, 89, 110
Ardennes, 27
Argentina, 35, 44, 66; in Falklands, 6, 8, 27, 28
Armalite, 161
Armenia, 64
Arnhem, 100
Asia, 41, 44
Assam, 68
Athens, 88
Atlanta, 95
Atlas Mts., 34
Attila, 89
Audo Range, 13
Australia, 34, 35, 36
avenues of approach, 25, 26–27
Aylwin, Patricio, 67
Ayodhya, 60
Azerbaijan, 64

Baath Party, 57, 66
Bad Hersfeld, 142–146
Baghdad, 11, 13, 26, 27, 57
Bagnold, R. A., 165
Balkans, 32
Baluchistan, 7
Bangkok, 62
Bangladesh, 44
Bannockburn, 90
Barre, Said, 54
barriadas, 147
base aero-terrestres, 100

Basra, 13, 27
Battle of Britain, 101
Battle of the Bulge, 103
Bay of Pigs, 99
Beaufort Castle, 28
Beijing, 64
Beirut, 15, 24, 26, 28
Bekaa Valley, 15, 26, 28
Belorussia, 64
Benguela Railway, 53
Berbera, 13
Berbers, 35, 54, 55
Berlin, 65
Beydene, 54
Bhutan, 60
Bhutto, Benazir, 59
Biafra, 44, 54
Black Prince, 90
blitzkrieg, 115–116
Bolivia, 67
Bolsheviks, 78, 84, 97, 140
Bombay, 60
Borneo, 62
Borodino, 92
Bosnia, 65
Boulding, Kenneth, 80
Boyne, 91
Brazil, 34, 36, 48, 66
Brecon Beacons, 18
Breitenfeld, 90
bricks, 161
bridges, 26
British Army, 2, 149–161
Brooke, Peter, 161
Buenos Aires, 18
Bulgaria, 58, 65
Burma, 48, 61
Burundi, 44, 48, 53
Busaco, 91
Bush, George, 67

Cairo, 55
Cambodia, 6, 44, 61, 64
Cannae, 88
Caribbean, 7, 68
Casbah, 98
caste, 61
Castro, Fidel, 79
cattenaccio, 78
Caucasus, 32, 53, 58, 64
cavalry, 90–91
Celts, 89
Central America, 5, 7, 21, 23, 29, 34, 37
Central Intelligence Agency (CIA), 66, 67, 99, 127
Cetewayo, 94
Chad, 48, 54, 55
Chalons, 89
Chamorro, Violeta, 67
Charles XII, 91
Chickamauga, 94
Chile, 35, 44, 48, 67
China, 36, 38; geopolitics, 41, 46, 60, 62, 63; People's Liberation Army, 116; at war with Vietnam, 6, 8, 21, 26, 27, 29
Churchill, John, 91
Clark Air Base, 63
Clausewitz, Karl von, 25, 117, 165
close terrain, 122
cocaine, 67
cocaleros, 67
Colombia, 48, 67
column, 108
concentration camps, 127
Congo, 34
contras, 48, 67
cordon and sweep, 127, 158, 159
core area, 49
Costa Rica, 21
counterinsurgency, 2, 23, 125–134
cover and concealment, 25, 29, 32
Cowpens, 92
Crecy, 90
Croatia, 65
Cromwell, Oliver, 91
crush zone, 49
CS gas, 159, 160
Cuba, 7, 44, 66
Custer, George, 93
Cyprus, 44, 58, 65
Czechs and Slovaks, 65

Damascus, 15, 26, 28
Dar el-Baidar, 15

Dar es Salaam, 53
Datis, 88
Dayan, Moshe, 24
Delhi, 69
Deng Xiaoping, 63
desert, 24, 34, 101, 102, 121
Desert Storm, 57, 58, 102, 117
Deutschtum, 65
Dien Bien Phu, 100
Diredawe, 26
Dublin, 78, 79, 96, 140

East Prussia, 66
East Timor, 44, 48, 62
Ecuador, 66
Edward III, 90
Egypt, 41, 44, 55
Ein Hilweh, 29
Ein Zehalta, 28
Ejercito Guerrillero de los Pobres, 6
El Alamein, 101
El Salvador, 6, 8, 21, 23, 44, 48, 67
Engels, Friedrich, 72
England, 73, 76
Epaminandos, 89
Eritrea, 44, 48, 54
Estonia, 64
Ethiopia, 8, 13, 26, 48, 54
Euphrates, 13, 35, 36
Europe, 35, 38, 41, 65–66

Fabian tactics, 21, 110
Falklands (Malvinas): geopolitics, 5, 6, 8, 18, 48; terrain, 27, 29, 48, 99
Fao Peninsula, 27
field manual, U.S. army, 113, 141
fields of fire, 25, 28–29
Fiji, 63
flavelas, 147
flying column, 126
foco insurreccional, 23, 79, 157
formations, 108–109
France, 41, 44, 48, 73, 76; in Algeria, 97–98; military conservatism, 106; revolution, 74, 166
Franco-Prussian War, 95–96
Frente Farabundo Marti de Liberacion Nacional (FMLN), 6, 67, 126
Friendship Pass, 27
Front de Liberation National (FLN), 97
Fulani, 54
Fuller, J.F.C., 113, 114

Gagauz, 64
Gallipoli, 98
Galtieri, Leopoldo, 8–9
Gandhi, Rajiv, 60
Ganges River, 60
geopolitics, 2, 31, 49, 69
Georgia, 64
geostrategy, 2, 31
Germany, 65
Ghana, 48
Giap, Vo Nguyen, 62, 100, 101
Gibraltar, 8
GIUK Gap, 151
Glabrio, 89
Glover Report, 157, 159, 161
Goering, Hermann, 101
Golan Heights, 56
Gorbachev, Mikhail, 60, 64
Grant, U. S., 94
Greece, 41, 65, 89
Grenada, 67, 99, 120
Guadalcanal, 98
Guam, 63
Guatemala, 67; geopolitics, 6, 8, 44, 48; terrain, 21, 23
Guderian, Heinz, 115
guerra prolongada, 23, 158
guerre revolutionnaire, 152
guerrillas, 2, 9, 11, 23, 27; tactics, 125–127; terrain, 129–134
Guevara, Che, 23, 79
Guinea Bissau, 44
Gulf, 7, 27, 35, 51, 57
guns, 90–93
Gustavus Adolphus, 90

Hafez, Assad, 57
Haiti, 68
hammer and anvil, 18
Hannibal, 27, 88, 89
Hanoi, 21, 26
Harrar, 26

Hastings, 89
Haughey, Charles, 151
Haussmann, Georges, 139, 140
Hawr al Hawizah, 27
hearts and minds, 126
Hechmatyar, Gulbuddin, 59
Heights of Abraham, 27
Henry V, 90
Hercegovina, 65
Herman (Arminus), 89
Hijaz, 34
Himalayas, 32
Hindu Kush, 7, 15, 18, 34, 59
Hindus, 60
Hitler, Adolf, 101, 103
homelands, 52
Honshu, 38
Hormuz, Straits of, 7
horsemen, 89–90
Hume, John, 159
Hungary, 41, 65
Huns, 89
Hun Sen, 62
Hussein, Saddam, 6, 57, 58
Hutu, 48, 53

Ia Drang, 100
Ilipa, 88, 110
Inchon, 98–99
India, 34, 41, 44, 48, 60–61
Indochina, 41
Indo-Gangetic Plain, 7, 38
Indonesia, 41, 44, 62
Indus River, 35, 60
Inkatha, 52
intelligence, 1, 159–160
internment, 156
Iran, 44, 48, 57, 58, 59; at war with Iraq, 5, 6, 8, 11, 13, 26, 27
Iraq, 41, 44, 48, 58–59; at war with Iran, 5, 6, 8, 11, 13, 26
Ireland, 76
Irish Republican Army (IRA), 2, 30, 97, 140, 149–161
Isandalwana, 93–94
Israel, 41, 44, 48, 55, 56, 116; in Lebanon, 6, 8, 15, 26, 27, 28, 29

Jackson, Andrew, 93
Jaffna, 61
Jamaica, 48
Japan, 36, 38, 62; jungle fighting, 116
Java, 33, 62
Jordan, 44
jungle, 23

Kabul, 5, 15, 26, 59
Kalahari, 34
Kanakas, 63
Kandahar, 5
Karachi, 60
Kasmiye, 26
Katanga, 44, 53
Kazakhstan, 64
Keegan, John, 164, 167
Kenya, 44, 53
key terrain, 8, 57, 122–123
KGB, 64
Khartoum, 54
Khe Sanh, 100
Khmer Rouge, 6, 44, 61–62
Khomeini, Ayatollah, 6, 59
Khorramshahr, 13
Khrushchev, Nikita, 7
Kikuyu, 53
Kim Il Sung, 63
King, Frank, 156
Kitson, Frank, 152, 155–157
Korea, 36, 38, 41, 44, 63
Kuomintang, 61
Kurds, 35, 44, 48, 57, 58–59
Kuril Islands, 63
Kursk, 102–103, 115
Kutuzov, Mikhail, 92
Kuwait, 57

Latvia, 64
Latvian riflemen, 76, 97
LAW, 146
Lawrence, T. E., 134
Lebanon, 41, 44, 48, 56, 59; 1982 war, 5, 6, 8, 15, 24, 27, 29
Lenin, Vladmir, 72, 96
Leonidas, 89
Leuctra, 89
Liberia, 53
Libya, 6, 8, 48, 54, 55
Liddell Hart, Basil, 114, 126

line, 108, 110
Lin Piao, 72
Litani River, 15, 26, 28
Lithuania, 64, 90
Little Big Horn, 93
London, 18
longbows, 90
Long March, 79
Long Range Desert Group, 109
long way, 158
low intensity warfare, 2
Luanda, 53
Luftwaffe, 97, 101, 103, 115

MacArthur, Douglas, 116
Macedonia, 65
Macedonians, 89
McColl, Robert, 72
MacMahon, Marie de, 96
McNeill, William, 163–164
McStiofain, Sean, 157, 158, 160
Madagascar, 41
Magyars, 65
Maharasthra, 60
Malaysia (Malaya), 37, 41, 61–62
Mali, 54
Manchuria, 21
Manteuffel, Hasso von, 27
Mao Tse-tung, 72, 126
maps, 1, 119–120
Marathon, 32, 88
March to the Sea, 95
Market Garden, 99–100
Marshall, S.L.A., 116, 167
Marston Moor, 91
Marx, Karl, 72, 75, 78
Mauritania, 6, 54
Mediterranean, 35, 88–89
Melanesia, 63
Mengistu Haile Mariam, 54
Mesopotamia, 5, 11
Mexico, 34
Mexico City, 147
Miami, 68
Micronesia, 63
Middle East, 55
military geography, 1
Miltiades, 88
Mindanao, 44, 62
missiles, 7
Mitla Pass, 101–102
Mobilization: on foot, 9, 32, 113–114; with railways, 94–96
Mobutu, Sese Seko, 53
Mogadishu, 13, 54
Moldavia, 64
Moltke, Helmuth von, 95, 110
Mongolia, 64
monsoon, 9, 21, 33
Montcalm, Louis, 91
Montgomery, Bernard, 1, 101
Morgan, Daniel, 91
Morgarten, 90
Morocco, 6, 8, 11, 27, 41
Moscow, 92
Motorman, 140, 156, 158
mountain terrain, 121
Mozambique, 34, 52
Mujahedin, 5, 18, 26, 28, 51

Nagas, 60
Nagorno Karabakh, 64
Najibullah, Mohammed, 59
Namib Desert, 34
Namibia, 52, 53
Napoleon, 91–92
Napoleon III, 95
Naseby, 91
Natal, 52
nationalism, 166–167
NATO, 151
Nazis, 77
Ndebele, 48, 58
Nelson, Horatio, 107
Nepal, 6
New Caledonia, 63
New Mexico, 34
New Orleans, 93
Nicaragua, 6, 8, 44, 48, 66–67
Nigeria, 44, 48, 54
Nile River, 33
"no-go" areas, 155–158
Noriega, Daniel, 67
Normandy, 98, 115, 116
Northern Ireland, 30, 85, 149–161
North Korea, 57
nostalgia, 166, 167

obstacles, 25, 27
Obwalden, 32
Oder-Neisse line, 66
Oder River, 90
Ogaden, 5, 6, 8, 13, 27, 48
oil, 7–8, 35, 55–56
oilspot, 126
Okinawa, 98
Oldfield, Maurice, 157
one-shot sniper, 161
open terrain, 122
Orinoco River, 34
Ortsinn, 25, 165

Paitak Pass, 26
Pakistan, 44, 59–60
Palestine Liberation Organization (PLO), 6, 15, 29
Panama Canal, 7
Panama City, 67, 98, 148
Panjshir Valley, 18, 28
Paraguay, 66
Paris Commune, 84, 96, 140
passes, 26
patria chica, 166
Paulus, Friedrich von, 97
Persians, 88, 89
Peru, 48, 67
Peter the Great, 91
Petrograd, 77, 97, 140
Philippines, 33, 41, 44, 62
Phnom Penh, 62
Pinochet, Augusto, 67
points of observation, 25, 27–28
Poitiers, 90
Poland, 65
pole weapons, 90
Polisarios, 6, 8, 11, 27, 41, 48, 55
Pol Pot, 62
Poltava, 91, 109
Polynesia, 63
Port au Prince, 68
Port Stanley, 6, 27, 28
Portugal, 44
Powell, Enoch, 151
power gradient, 49, 50, 80–85
Provos, 152–161
psycho-social targets, 155
Punjab, 59–60

Qaddafi, Muammar, 55
Quebec, 68, 91
Quick, John, 141

railways, and mobilization, 94–96
rain forest, 23, 33
Ramillies, 91
ratio of fire, 167
Red Army, 97, 115
Red River, 21
Republic of Ireland, 151
revolution, 71–85
Reynoso, Abimael Guzman, 138
Rezendes, Scott, 129
Rhodesia (Zimbabwe), 44
Richard I (Lionheart), 109
ring of steel, 156
Rios-Montt, Efrain, 23, 68
Riyadh, 35
Robert the Bruce, 90
Rodophe Mts., 65
Romania, 64, 65
Romans, 89
Rommell, Erwin, 24, 101
Rourke's Drift, 94
Royal Air Force (RAF), 100
Royal Ulster Constabulary (RUC), 149, 150, 157
Rundstedt, Gerd von, 27
Russia, 64, 65, 76, 107
Rwanda, 44, 53

Sahara, 9, 31, 34
St. Petersburg (Petrograd, Leningrad), 77, 97, 140
Salan, Raoul, 100
Salang Pass, 5, 26
San Carlos, 27, 29, 99
Sandinistas, 5–6, 44
San Salvador, 148
savanna, 24, 34
Schweinfurt, 142–146
Scipio Africanus, 88

Scotland, 76
search and destroy, 126
see-saw strategy, 158
Sendero Luminosa, 138
Sepoy Mutiny, 76
Sheba (Katanga), 53
Shah, 7, 44, 59, 75
Shaka, 93, 110
shantytowns, 147
Sharon, Ariel, 15, 24
Shatt al Arab, 6, 8, 11, 13, 27, 29, 59
Shemlan Ridge, 26, 28
Sherman, William, 94–95
Shona, 48, 52
Shouf Mts., 15, 28
Sidon, 24, 29
Sierra Maestra, 79
sight lines, 9, 24, 114; in cities, 141–146
Sinai Peninsula, 101, 102
Sind, 60
Singapore, 62
Sinn Fein, 159, 161
Six Day War, 102
Slim, William, 116
snow, 121–122
"softly, softly," 155
Somalia, 6, 8, 13, 27, 44, 54
South Africa, 48, 51–52, 81
South Yemen, 48
Spain, 6
Sparta, 89
spacial ability, 164–166
Special Air Services (SAS), 157, 159, 160
sphere of influence, 50
square, 108–110
Sri Lanka, 44, 48, 61
Stalingrad, 97, 115
steppe, 36
Stinger missile, 15, 27, 116, 161
strategy, 1, 2, 105
Sudan, 44, 48, 54
Suharto, 62
Sun Tzu, 126, 159
Sun Yat Sen, 78
Swiss, 90
Syria, 6, 8, 15, 27–29, 44, 48, 56–58
tactics, 1–2, 31, 105–111
taiga, 37
Taiwan, 38, 41, 63–64
Tamils, 48, 61
tanks: limitations, 119; in towns, 146
Tannenberg, 90
Tanzania, 53
Tauregs, 54
Tbilisi, 64
Teheran, 11, 59, 75
terrain analysis, 24–29, 31, 128–129
territory, 1, 49, 50, 162, 166
Teutobergerwald, 89
Teutonic Knights, 90
Thailand, 61
Thatcher, Margaret, 8–9
Thermopylae, 32, 88–89
Third World, 72
Third World cities, 147–148
Thrace, 65
Tiananmen Square, 63–64
Tiber River, 60, 64
Tigre, 54
Tigris River, 13, 35
Timisaoro, 65
tonton macoute, 68
Torres Vedras, 91
TOW, 119, 146
Transvaal, 52
Transylvania, 65
Trasimene Lake, 89
trench warfare, 94, 114
Trinidad, 68
Tukhachevski, Mikhail, 114
Tunisia, 41, 55
Turkey, 44, 58–59, 65
Tutsi, 48, 53
Tuzo, Harry, 156
Tyre, 2, 24, 26

Uganda, 44, 53
ugdah, 15, 102, 116
Ukraine, 64
Ulster Defense Regiment (UDR), 157, 158
Ulundi, 94
U.K., 66; in Falklands, 6, 8, 18, 27,

28, 30; in Northern Ireland, 149–161
U.S., 7, 8; Army doctrine, 140; geopolitics, 36–38, 41, 44, 53, 57–59, 66–67
U.S.S.R.: in Afghanistan, 5, 7, 8, 18, 21, 28; geopolitics, 41, 44, 48, 50, 53, 58–59, 64–65
urban: spine, 147–148; warfare, 2
Uttar Pradesh, 60

Valley Forge, 78
Varus, 89
Vicksburg, 94
Vietnam: against China, 5, 6, 8, 21, 26, 27, 29; geopolitics, 44, 61–62, 64
Vlachs, 65

wadis, 11, 27
Wales, 76
Wallace, Mike, 127
Waterloo, 92
Watts, 68
Wehrmacht, 97, 103, 115–117
Wellington, Duke of, 91–92, 106, 109
Western Sahara, 5, 6, 9, 27, 55
West Irian, 62
Westmoreland, William, 100, 127
West Sumatra, 62
Wheatcroft, Andrew, 73
whirlwind, 126
William I (the Conqueror), 89
William III (of Orange), 91
Wingate, Orde, 100
Wolfe, James, 27, 91
Wolog, 54
Wordsworth, William, 74
World War II, 37

Xerxes, 89
Xhosa, 48, 52
Xuzhou, 103

Yangon (Rangoon), 6, 139–140
Yemen, 44, 58
Yom Kippur War, 102
yomping, 18
Yugoslavia, 65
Yunnan-Guangxi, 18

Zagreb, 65
Zagros Mts., 11, 13, 26, 34
Zaire, 44, 53
Zambezi River, 34
Zambia, 44, 53
Zanzibar, 54

About the Author

PATRICK O'SULLIVAN is a Professor and Chair of the Department of Geography at Florida State University. He is the author of numerous books, articles, and book chapters, including a previous study of the geography of warfare.